Workplace Involvement in Technological Innovation in the European Community

Volume II:

Issues of Participation

EF/93/04/EN

Title Series

Volume I   Roads to Participation
           ISBN 92-826-5669-1   Cat. No. SY-17-93-001-EN-C
           ECU 31,50
           (Revised report originally produced as research report
           Cat. No. SY-72-91-035-EN-C   ISBN 92-826-3186-9

Volume II Issues of Participation
           ISBN 92-826-5672-1   Cat. No. SY-17-93-002-EN-C
           ECU 31,50

Workplace Involvement in Technological Innovation in the European Community

Volume II:

Issues of Participation

Colin Gill
Thérèse Beaupain
Dieter Fröhlich
and
Hubert Krieger

European Foundation for the Improvement
of Living and Working Conditions
Loughlinstown House, Shankill, Co. Dublin Ireland.
Tel: +353 1 282 6888 Fax: +353 1 282 6456 Telex: 30726 EURF EI

Cataloguing data can be found at the end of this publication

Luxembourg: Office for Official Publications of the European Communities, 1993

ISBN 92-826-5672-1

© European Foundation for the Improvement of Living and Working Conditions, 1993.

For rights of translation or reproduction, applications should be made to the Director, European Foundation for the Improvement of Living and Working Conditions, Loughlinstown House, Shankill, Co. Dublin, Ireland.

Printed in Ireland

Preface

The 1980s have seen a widespread debate on workplace involvement in the process of technological innovation in Europe. The two volumes of this report present the results of an attitudinal survey on the impact of new information technology in enterprises across the 12 Member States of the European Community and outline and compare the experience and opinions of nearly 7,300 managers and employee representatives.

The introduction of new information technology in European enterprises during the 1980s has had profound implications both for employees and the way such enterprises have been managed. Participation in technological change has been accorded a high priority by the Social Partners and thus the survey gives an unprecedented opportunity not only to evaluate the effects of these changes stemming from new technology, but also to give some detailed information on the degree of participation by employee representatives in such technological change.

The **first volume** of the report "**Roads to participation**" is focused on participation during the different phases of the process of technological innovation. The introduction of new technology can be viewed as going through a number of phases: planning, selection, implementation and post-evaluation. The first two phases are strategic, the latter two are operational.

The **second volume** of the report "**Issues of participation**" produces valuable results on the intensity of participation on different issues directly related to technological innovation. In particular, the survey covers participation in training, health and safety aspects of new information technology, work organisation, investment, product quality and the benefits of participation to both managers and employee representatives. It also gives an overview about the perceptions of the respondents regarding the impact of new information technology on employment levels, particular occupations, employment flows within enterprises, the problems that new information technology poses for particular occupational categories, and the role of participation in dealing with these problems.

Perhaps the most important conclusion that emerges from the survey is the sheer diversity in the forms of participation that exists in the individual countries. Taking all explanatory factors together, it is clear that the types of participation that exist in Europe, or which will emerge in the future, are a product of the way that each Member State's industrial relations system has been shaped by wider political, economic, social and historical forces.

European Foundation for the Improvement of Living and Working Conditions	European Foundation for the Improvement of Living and Working Conditions
Clive Purkiss	Eric Verborgh
Director	*Deputy Director*

Table of Contents

SUMMARY 1

CHAPTER ONE: INTRODUCTION 17
1.1 Introduction 18
1.2 Major research questions of the survey 18
    1.2.1 The employment impact of new information technology 20
    1.2.2 Participation in training 20
    1.2.3 Participation in health and safety 21
    1.2.4 Participation in work organisation 21
    1.2.5 Participation in investment in new information technology 22
    1.2.6 Participation in product quality 22
    1.2.7 The benefits of participation to both sides 22

CHAPTER TWO: SURVEY METHODOLOGY 25
1. Countries and Sectors 26
2. Survey Procedure 27
3. Limitations of the study 28

CHAPTER THREE: THE EUROPEAN POLITICAL CONTEXT .. 31
3.1 Introduction 32
3.2 The Social Dialogue 32
3.3 Proposals of the European Commission for worker participation 34
3.4 The Community Charter of Fundamental Social Rights for Workers 36
3.5 The views of the Social Partners 38
3.6 Discussion 40

CHAPTER FOUR: DEVELOPING AN EXPLANATORY FRAMEWORK FOR PARTICIPATION IN TECHNOLOGICAL CHANGE AT COUNTRY LEVEL 43
4.1 Introduction 44
4.2 Five Principal Factors 44
    4.2.1 Management's reliance on its workforce to achieve its objectives 45
    4.2.2 Management Style and attitude towards participation 47
    4.2.3 The bargaining power of organised labour 48
    4.2.4 Legal Regulation 51

      4.2.5 The degree of centralisation of the industrial relations system 52
4.3 Conclusions 53
      Table 1. Factors affecting employee participation in technological change 54
      Table 2. Country-specific factors and their influence on participation levels in particular Member States ... 55

CHAPTER FIVE: THE EMPLOYMENT IMPACT OF NEW INFORMATION TECHNOLOGY 57
5.1 Introduction 58
5.2 Employment effects at the level of the firm at European level: separate effects 59
5.3 Analysing the employment effects with greater precision ... 61
5.4 Combined employment effects at country level 63
5.5 The quantitative employment effects of new information technology 64
5.6 Occupations affected by technological change 68
5.7 The role of participation in ameliorating the effects of new technology on particular groups — European level 70
      5.7.1 Middle managers 71
      5.7.2 Office staff 72
      5.7.3 Secretarial staff 72
      5.7.4 Shopfloor skilled workers 73
      5.7.5 Shopfloor semi-skilled and unskilled workers 73
5.8 The role of participation in ameliorating the effects of new information technology on particular groups — country comparisons 75
      5.8.1 Middle managers 75
      5.8.2 Office staff 77
      5.8.3 Secretarial staff 78
      5.8.4 Shopfloor skilled workers 80
      5.8.5 Shopfloor semi-skilled and unskilled workers 81
5.9 General 83
5.10 Conclusions 83

CHAPTER SIX: PARTICIPATION IN TRAINING 87
6.1 Introduction 88
6.2 Past participation in training at the European level 89
6.3 Inter-country comparisons of participation in training 90
6.4 Explaining inter-country differences in training 92
6.5 Participation in various training issues 97
6.6 Inter-country comparisons of participation in various training issues 98

      6.6.1 Selection of participants........................ 98
      6.6.2 Timing and duration............................ 99
      6.6.3 Contents of training........................... 100
      6.6.4 Type of training.............................. 100
6.7 Assessing the potential for participation in training in the future 101
6.8 Assessing the results in the light of the Val Duchesse Joint Opinion 104
6.9 Conclusions 106

CHAPTER SEVEN: PARTICIPATION IN HEALTH AND SAFETY 109
7.1 Health and safety at the European Community level....... 110
7.2 Health and safety within the Member States.............. 113
      Table 3: Institutional arrangements and information and consultation rights in health and safety.......... 116
      Table 4: Workers' representatives' rights in health and safety 117
7.3 Survey results... 117
7.4 Conclusions 123

CHAPTER EIGHT: PARTICIPATION IN WORK ORGANISATION 127
8.1 Introduction... 128
8.2 Participation at European level........................ 130
8.3 Participation at Member State level.................... 131
8.4 Conclusions 139

CHAPTER NINE: PARTICIPATION IN INVESTMENT IN NEW INFORMATION TECHNOLOGY........................... 141
9.1 Introduction... 142
9.2 Participation in investment at the European level......... 144
9.3 Participation at Member State level..................... 144
9.4 Future participation in investment in new information technology... 147
9.5 Assessing the results in the light of the Val Duchesse Joint Opinion 151
9.6 Conclusions 152

CHAPTER TEN: PARTICIPATION IN PRODUCT QUALITY .. 155
10.1 Introduction.. 156
10.2 Past participation in product and service quality at European level 158
10.3 Comparative results in the 12 Member States............. 159
10.4 Future expectations of participation in product and service quality 161

10.5 Assessing the results in the light of the Val Duchesse Social Dialogue... 164
10.6 Conclusions 164

CHAPTER ELEVEN: THE BENEFITS OF PARTICIPATION.... 167
11.1 Introduction....................................... 168
11.2 Assessing the survey results.......................... 168
11.3 Mutual understanding in the firm..................... 170
11.4 Utilisation of knowledge and skills.................... 173
11.5 Decision-making.................................... 177
11.6 Implementation of new information technology......... 181
11.7 Conclusions 183
    Table 5 185
    Table 6 186

CHAPTER TWELVE: CONCLUSIONS...................... 187

Summary

1. Introduction

This report presents the results of an attitudinal survey on the impact of new information technology in enterprises across the 12 Member States of the European Community. It outlines the experience and opinions of nearly 7,300 managers and employee representatives from the countries of the Community. The survey, which was carried out by the **European Foundation for the Improvement of Living and Working Conditions** covered five important industrial sectors; mechanical engineering and electronics in the manufacturing sector, and banking, finance and retailing in the services sector. The report concentrates on examining in some detail the impact that new information technology has had on jobs in European companies and the kind of participation that accompanies new information technology when it is introduced.

The study is essentially comparative and the core questioning in the survey produced much valuable information about the perceptions of the respondents on the impact of new information technology on employment levels, particular occupations, employment flows within enterprises, the problems that new information technology posed for particular occupational categories, and the role of participation in dealing with these problems. A central focus of the study concerned the role of participation in technological change; in particular, the survey covered participation in training, health and safety aspects of new information technology, work organisation, investment, product quality and the benefits of participation to both managers and employee representatives.

The introduction of new information technology in European enterprises during the 1980s has had profound implications both for employees and the way such enterprises are managed. Participation in technological change has been accorded a high priority by the Social Partners and thus the survey gives an unprecedented opportunity not only to evaluate the effects of these changes stemming from new technology, but also to give some detailed information on the degree of participation by employee representatives in such technological change.

2. Survey Methodology

The survey was designed to measure opinion of managers and employee representatives who were personally involved in the introduction of new information technology which had a significant impact in the workplace. The methodology was agreed upon by a group of experts including representatives of unions and employers at a European level.

In the analysis of technology introduction and its results equal weight was given to the information provided by, and opinions of, both managers and the employee representatives. This was achieved by ensuring that pairs of managers and employee representatives of the same company or establishment were interviewed. The survey sample was restricted to larger establishments (at least 50 employees) which had some kind of formal employee representation and which involved these representatives in the process of technology introduction. The response rate was around 50% and, as is common in surveys of this kind, there was inevitably a degree of self-selection inherent in those companies which took part in the survey.

Altogether, a total of 7,326 interviews — with 3,663 managers and the same number of employee representatives — were carried out in the 12 countries of the European Community. The interviews were carried out in two waves: between 11 February and 22 May 1987 in Denmark, Germany, United Kingdom, France and Italy; in Belgium, Greece, Ireland, Luxembourg, Netherlands, Portugal and Spain between 5 April and 7 October 1988. Altogether, interviewing took place in 2,807 European companies which means that — as a statistical average — 1.3 managers and 1.3 employee representatives were interviewed in each company. This is the **largest** survey to be conducted on the topic of employee representatives' involvement in technology introduction in Europe.

3. The Political Context

Many of the issues which arise in this report have to be seen within the context of political changes which are currently taking place as the European Community moves towards the setting up of the Single Market in 1992. An important feature of the move towards the Single Market is what has been called the Social Dimension. There are a number of features of the Social Dimension which are relevant to the results of our survey.

The first of these is the **Social Dialogue** between the two sides of industry at European level (the European Trades Union Confederation, ETUC, on the one hand, and the Union of Industries of the European Community, UNICE, and the European Centre of Public Enterprises, CEEP, on the other), which was reactivated by Jacques Delors at the beginning of 1985. This type of Dialogue is covered by the Single European Act, Article 118B and these discussions have led to a number of **Joint Opinions** on information and consultation as to the introduction of new information technologies in firms. Recently, the Social Dialogue has been seen as the main means of implementing many of the measures in the **Social Action programme** to put into effect the **Community Charter of Fundamental Social Rights for Workers**. Directives will only be issued in circumstances where the Social Dialogue process is seen as inappropriate for introducing such measures.

Secondly, in the last two decades the European Commission has put forward several proposals over the years for worker participation on the grounds that democratisation of company management can not only be a factor promoting industrial peace but also one which improves the quality of management itself. These range from the so-called **Fifth Directive**, the **Vredeling Directive,** and the **European Company Statute**. However, these attempts to regulate and extend workforce participation in Europe by institutional and legal forms of intervention have not been successful, at least for the present. The most recent proposal concerned with participation within European companies was the draft **Directive on European Works Councils** on the procedures for the information and consultation of workers of European-scale undertakings.

Thirdly, the European Commission has also issued a number of Directives concerned with **health and safety at the workplace**. Among them is the **Framework Directive** adopted in June 1989 on the introduction of measures to encourage improvements in the health and safety of workers at the workplace. This Directive establishes the framework for the new European Community policy on workers' health and safety. Its scope is very wide, covering both private and public sectors and laying down a number of general obligations to be met by the employer. Apart from its health and safety provisions, the Directive also deals with the disclosure of information to workers, means of consultation, and provisions for participation and training in health and safety matters.

Finally, the Commission has proposed a number of other Directives concerning **information and consultation** — all of which have already been put into practice. These include the **Collective Redundancies Directive** on procedures to be applied in redundancy situations, which was adopted in 1975; and the **Transfer of Undertakings Directive**, which provided for employees' rights and advantages in mergers and take-overs, which was adopted in 1977. The Commission has also published a **Recommendation on equity-sharing and financial participation by workers in enterprises** on 12 July 1991 as part of its work programme for 1991 on the **Social Charter**.

The Social Charter was given added importance by the extension of qualified majority voting to a wide range of social and employment issues by the 11 Member States which ratified the Social Chapter of the Maastricht treaty in December 1991. Our results can act as a useful benchmark in assessing current progress in the Action Programme and in pointing to areas which need attention in the move towards greater involvement by employees in the European organisations in which they work.

4. The Views of the Social Partners

Whilst the Social Dimension of the 1992 Single Market is now receiving much attention in the Community as a whole, there is a wide divergence of views as to how the various strands of the social dimension should take effect. Both sides of industry have their own specific interests and policies on the social dimension of the Single Market. These divergent views are reflected in the differing positions taken on participation in technological change. Whilst there is a small degree of consent between all parties, there is also a great deal of disagreement.

The ETUC believes that negotiation should take place with management at all stages concerned with the introduction of new technology into the workplace.

The ETUC position is strongly opposed by UNICE. UNICE insists on the right of company management to take final decisions. It further rejects any form of statutory regulation at the European level:

> "Employers do not oppose practices such as co-determination, let alone information and consultation. They strongly oppose the imposition of such practices on all Member States by means of EC-level legislation. Such matters must be worked out by the parties most directly concerned, according to local conditions, traditions and legislation".

5. An Explanatory Framework to interpret our results

In the Report we set out an explanatory framework which is designed to explain the different levels of participation that existed throughout the 12 Member States at the time the interviewing took place. This explanatory framework consists of a number of **factors** which assist or impede the possibilities of employee involvement in the introduction of new information technology and which are used to explain the survey data. There are a number of country-specific factors which shape the opportunities for employee involvement in the introduction of new information technology. It is important to mention that some of these factors will not only be country-specific but will also apply at sectoral level and company level in the countries themselves. The factors will also be interrelated to some degree, and strong forms of particular factors might well have a dominating impact on the outcome of the degree of employee participation in technological change that emerges. These factors will not only shape the opportunities for participation but they will also affect the subject matter and timing of involvement. There are five principal factors which can help or hinder the opportunities employee representatives have for participation; they will not, however, determine the extent to which employee representatives can influence the effects of technological changes in the organisation. It is one thing to put discussion about technological change on to the formal agenda and quite another to translate these participative opportunities into a real ability to influence the way new information technology can affect working life. The factors used in the explanatory framework are:

> **management's dependence on the skills and co-operation of its workforce to achieve its objectives for introducing the new information technology;**
>
> **management style and its attitude to participation;**
>
> **the bargaining power of organised labour to force management to negotiate or consult with its representatives in the absence of any voluntary disposition on the part of management to do so;**
>
> **regulations which lay down participation rights for employees or their representatives on a range of matters at enterprise level;**
>
> **the degree of centralisation of the industrial relations system that exists in the country concerned;**

Each of these country-specific factors will have a number of favourable and unfavourable conditions attached to them, depending on the country concerned.

6. The employment impact of new information technology

New information technology has had a beneficial effect on the jobs of those employees in the forefront of technological change. When it has been introduced it has encountered little resistance and few employees express fears about its adverse effects. There is little evidence in our survey results to suggest that jobs have disappeared as a result of the introduction of new information technology into European workplaces; on the contrary, what evidence we do have points to the opposite conclusion, namely that if anything, more jobs have been created as a result.

It also seems that participation provides an effective means of resolving the problems that employees in various occupational groups may have about the impact of new information technology on their jobs. This result is by no means surprising but it lends support to those who believe that such participatory machinery is essential in the management of change in companies.

One of the central findings of this part of the survey — is that new information technology appears to have created jobs rather than destroying them, at least in the sectors in which the survey was carried out.

New information technology has also had a pervasive, if generally beneficial effect, on jobs throughout European enterprises during the 1980s where large numbers of employees from all occupational groups felt its effects. In this respect, the early predictions that were made about new information technology being a truly "heartland" technology have certainly been realised. It seems from our results that new information technology tends to have a large impact at the top of organisations, with large numbers of managers and office staff affected.

7. Participation in training

It is clear that training has a fundamental role to play in strengthening the competitive position of European firms and in reviving growth. The skills and competences of the workforce, from production workers to senior management, are proving to be crucial determinants of productivity and competitiveness. There is a great deal of consensus between the two sides of industry about the importance of training. Given this consensus, the success of training in the future depends on a partnership not only between governments, but also between management and employees.

It seems from our results that there is no guarantee that participation in training will occur when new information technology is introduced into the workplace. Given that our sample covered only those firms which already had some form of institutionalised representation, one would expect that our data would indicate a much greater degree of partnership between managers and employee representatives in training matters than it does.

Only half of the employee representatives were either consulted about training or had a stronger form of involvement in this important area. More ominously, given the importance that training occupies on the agenda of both managers and employees in the 1990s, it is particularly chilling to note that **one out of every five employee representatives have no involvement in training whatsoever** and in six of the 12 Member States less than half of employee representatives were consulted about training.

However, the survey does provide evidence that all countries of the Community are moving in different degrees towards a greater level of participation in the future. Moreover, at European level, there is a high measure of agreement between the two sides of industry in relation to the Val Duchesse Social Dialogue.

Two of the key elements of Community policy in education and training — producing a better qualified workforce and ensuring a greater equality of opportunity between regions — might have a greater chance of success if our survey results showing greater levels of co-operation between the two sides of industry in the future prove to be correct.

8. Participation in health and safety

Health and safety now occupies a prominent place on the European political agenda. This appears clearly in article 118A of the Single European Act, in the Charter of Fundamental Social Rights as well as in the Action Programmes and the Directives which have been issued. Among the latter, the Framework Directive on the introduction of measures to encourage improvements in the health and safety of workers at the workplace is of particular interest for this survey.

Health and safety appears to foster a comparatively higher level of participation in the process of introducing new technologies. Nevertheless, wide differences exist between the countries of the European Community, both in practice and in future expectations.

There are only two countries where at least 70% of the firms respect the consultation objective of the Framework Directive on health and safety — **Germany** and **Denmark**. **Belgium, Ireland, United Kingdom** and **Netherlands** are still above 60%. **Luxembourg** and the southern countries of Europe, including **France**, are in a worse position, especially **Portugal**, where consultation and higher forms of participation are not even practised in 5% of the firms.

Nevertheless, our results indicate that the deficit between employee participation in practice and the Framework Directive's minimum requirement will markedly reduce within a foreseeable period of time. Only 24% of managers and 11% of employee representatives in Europe are opposed to a satisfactory practice of consultation within the near future.

9. Participation in work organisation

One of the major concerns of employee representatives about the introduction of new information technology into their organisations is the danger that it will be accompanied by forms of job design and a division of labour which are based on "Taylorist" principles of "scientific management". Indeed, participation in work organisation is central to trade union interests.

There is a great deal of variation across the European Community in the premium that is placed by trade unions in different Member States on the importance of good work organisation. In some countries, notably in **Denmark, Germany** and to a lesser extent the **Netherlands**, unions have traditionally taken a strong interest in job design. Trade unions in other EC countries have not. We would expect our survey results to show a high degree of participation in work organisation in both these two countries. Both countries have well-resourced trade union movements, well-trained union officials, well-established traditions of concern for working life initiatives from both sides, management traditions that are not openly averse to issues of work organisation and some degree of governmental support. In contrast, we would expect countries with trade union movements which are either divided along religious or political lines and/or which are poorly resourced to place less attention to QWL issues; **Portugal** and **Greece** are examples.

Our survey results reflect the North-South contrast in the importance given to participation in work organisation. In the northern European countries such as **Denmark, Germany**, the **Netherlands** and **Belgium** work organisation assumes much more importance than is the case in the Mediterranean countries, where little priority is given to work organisation.

There is a cluster of countries for which no involvement is low and co-determination is high: **Germany, Denmark** and **Ireland**. The absence of involvement is also low but the amount of co-determination less prominent in the **Netherlands, Belgium** and the **United Kingdom. Luxembourg, France, Spain, Greece** and most noticeably, **Portugal** have much lower levels of participation in work organisation.

The expressed preferences of both sides of industry in this issue give some hope for enhanced participation levels in these matters in the future. Some degree of involvement is expected in all Member States, except in **Portugal**, and to a much lesser extent, in **Luxembourg**.

Participation is largely meaningless unless it is about work organisation; it is central to all of the trade union concerns.

10. Participation in investment in new information technology

During the late 1970s there was growing concern within the trade union movement in many European countries with the issues that were raised by new computing and information technologies. Unions emphasised in their policy statements that if the adverse implications of the new information technology were to be avoided, then unions would have to seek more effective ways of negotiating its introduction.

The ETUC emphasise that procedures for change should be based on collective bargaining and joint decision-making whose progress would be monitored by joint bodies; consultation with trade unions should begin prior to the decision to introduce new information technology; information and management plans should be made available to the unions and until such time as agreement was reached, the status quo should be observed.

Because participation in investment strategy is concerned primarily with the **planning** phase of technological change, we would not expect very high levels of participation since management sees this aspect of technological change as essentially their prerogative.

Denmark and **Germany** have the highest levels of negotiation/joint decision-making among the twelve European countries, respectively 15% and 16%. The **Netherlands** comes next with 55% of managers stating that there is at least information of employee representatives on investment strategy in their firms. This percentage falls below 50% for the nine remaining countries in the Community.

The countries where participation in the investment criteria of new technology falls to its lowest levels — less than one respondent out of four — are **Greece**, the **United Kingdom** and **Portugal**.

In **every** country managers envisage higher levels of participation in the future. This trend towards higher levels of participation is particularly marked in the **Netherlands**, **Denmark** and **Germany**.

It is clear from our survey data on participation in investment criteria for new information technology that the levels of participation are very low indeed. There is nothing surprising about these results because this aspect of technological change in enterprises is concerned with the **planning** phase, and management has always seen decisions of this kind as their prerogative.

Despite the fact that the ETUC policy is to ensure that all aspects of technological change are the subject of negotiation or joint decision-making, their success in achieving this objective has so far been minimal. There is evidence that management is prepared to take a more accommodating view about involving employee representatives in investment decisions in the future.

Our survey results also reflect the general trend throughout Europe for management to seek much more decentralisation and de-regulation in industrial relations. In many enterprises strategic decisions about investment in new technology are taken at higher levels in the company which are far removed from the establishment itself and this severely limits the scope of trade union influence at the level of the establishment. Management is placing emphasis on employee involvement in the **implementation** phase of technical change.

Compared with other aspects of participation in new information technology such as training, health and safety, and work organisation, our survey results show that levels of participation in investment decisions concerned with technological change are very low indeed.

11. Participation in product and service quality

One of the primary reasons for the creation of the Single Market in the European Community was to enable European enterprises to meet the challenge of competition in world markets from countries such as Japan and some nations in the Far East; the quality of products and services is therefore of prime importance for the success of European enterprises.

Management in many European enterprises have adopted several aspects of Japanese managerial practice and this has led to an emphasis on the links between greater status equality, labour flexibility and a sense of personal commitment to the quality of finished products and services. This change in managerial thinking is itself linked to two strands of managerial thinking which gained prominence during the 1980s; total quality management and human resource management.

Management in Europe tends to restrict employee representative involvement merely to the provision of information about product and service quality, and there were numerous examples of no involvement by employee representatives whatsoever.

Several countries stand out with favourable results in this area: **Denmark, Greece, Ireland,** the **Netherlands,** and the **United Kingdom.** The results for **Luxembourg** and **Portugal** were particularly poor. The traditional ideological reluctance of trade unions to be involved in what they see as being party to unpopular management decisions is also reflected in our survey results for **France** and **Italy;** however, there is also evidence that this is changing, with both management and employee representatives in those two countries expressing a willingness to see product and service quality as an issue where both sides can make a constructive contribution to the success of French and Italian enterprises.

Given the importance of this issue for the competitiveness of European enterprises in world markets, our results show that there is likely to be much stronger forms of participation in the years to come. In all countries there is a shift towards stronger forms of participation, but more so in some countries than others. In **Denmark, Greece, Ireland,** the **Netherlands,** and the **United Kingdom** well over three out of every four managers indicate that future participation in product and service quality will be characterised by consultation and co-determination. The shift towards higher levels of participation is not so pronounced in **Belgium, Germany, Luxembourg** and **Portugal.**

Generally, our survey results indicate that future participation levels in Europe in product and service quality are likely to be healthy. Both sides of industry appear to be aware of the importance of this issue for enhancing the competitiveness and productivity of European enterprises.

12. The benefits of participation

The survey results show that in strategic topics (such as investment decisions about new technology), the more binding forms of involvement such as consultation, negotiation and joint decision-making were fairly rare. These higher levels of participation were more prominent in regard of direct workforce concerns such as health and safety, work organisation and training. Both sides want more participation in the future.

Nevertheless, important differences remain. Management prefer for the future increased information and consultation, whereas employee representatives tend to seek negotiations and joint decision-making.

The overall impression which we gain of participation practices in high tech companies in the European Community is one of ambivalence: in strategic company matters the situation can be described as a combination of no involvement at all or low intensity of involvement of workforce representatives; whereas in operational problems there is a clear trend towards more intensive participation practices.

What are the effects of participation on the technical modernisation of European firms? The overall situation can be characterised as a mixture of no impact and positive results. When managers and employee representatives indicated effects they were mainly, **sometimes overwhelmingly positive**. Participation does not slow down managerial decision-making nor does it delay the implementation of new technology systems; there is clear evidence that it enhances mutual understanding between managers and employees in companies and leads to an improved knowledge and utilisation of skills on the part of employees. It leads to a better industrial relations climate and does not adversely affect the quality of managerial decision-making.

A comparison between the 12 Member States shows important differences in the assessment of the benefits of participation. Portuguese managers are least positive in all five dimensions. Their general perspective is that participation has no impact on the economic and social performance of the company. This general attitude is also taken by the Portuguese employee representatives. **It is important to note that in regard to the benefits of participation no joint pattern for the peripheral countries of the EC can be observed. Portugal** is the exception. The reason for this may be the immaturity of the Portuguese system of industrial relations, the short democratic tradition and the recently introduced legislation in this area. But these arguments are not fully convincing as they **also** partly apply to **Spain** and **Greece**.

An even more astonishing feature is the negative attitude of German managers. In regard to the key criteria for the valuation of participation, German management is significantly more pessimistic than the European average on the management side. The German results are even more exceptional, if one compares them with the Danish, Dutch and Belgian results. **Denmark** and **Germany** are the most advanced countries with regard to participatory practice in the EC. Within Danish management two out of five categories are significantly more positive than the European average. From a previous study, we know that there is a statistically significant relationship between the strength of participation and its positive evaluation; but this is **not** true for German management. The reason for this may be tactical reasoning, i.e. that a positive response may be used against German management to reinforce legislation.

One also can compare the attitudes of Danish and German managers on another level. Previous research showed that Danish managers were much more open to strong forms of participation in strategic areas for the future than German managers. It seems to be that Danish managers, due to their positive experience with participation, are much more convinced about its value, whereas German management seems to be pushed into participation by legal regulation and strong unions.

Another interesting observation is that there is no relation between strong legal regulations and negative management perception of benefits. The last interesting result is related to the extraordinary positive assessment of British management in all five dimensions. An explanation may be the selection of UK companies. The survey sample consisted of companies with formal employee representation and some minimal practice in participation, i.e. considering the hostile political environment for participation in 1987, the companies in the survey were those which were prepared to develop participation without legal or contractual obligation and against a hostile political environment. It is not unreasonable to assume that it needs a strong conviction of management based on positive impacts of participation to develop some participatory practice under these circumstances.

The enthusiasm of British management **is not shown by British employee representatives**. Their perception does not differ from their European colleagues with the exception of one dimension. The most negative results can be found in **Portugal** and **Luxembourg**. The perceptions of employee representatives in **Luxembourg** is in line with those of German management. A system of co-operative industrial relations produces doubts on the benefit of participation. The most positive employee representatives in Europe can be found in **Ireland**. This is another indication that Irish industrial relations is increasingly diverging from the British tradition. Irish workers representatives and unions are much more positive about participation than their UK counterparts. On the management side it is vice versa. The German and Danish workers' representatives show more positives attitudes only in one dimension compared with their European colleagues.

13. Conclusions

This is the **only** survey which covers participation in technological change throughout the EC and therefore its results are particularly important to all the Social Partners at a time when the Community is moving towards greater economic and political union and when various social policy initiatives are being proposed involving information and consultation of employees.

Perhaps the most important finding is that where participation occurs, regardless of the issues involved, it is overwhelmingly seen by both sides as **beneficial and positive**. It is seen as conducive towards promoting better mutual understanding between management and employees in companies, a greater utilisation of knowledge and skills of employees in the process of modernising companies, a smoother process of technological implementation, and either a neutral or positive effect on the quality of decision-making. In general, our results point to a high degree of trust between both sides of industry.

There is evidence that managers and employee representatives are both seeking higher levels of participation in technological change in the

SUMMARY

immediate future. Both sides value the contribution that can be made by each other to promoting a smoother transition towards the modernisation of the enterprises in which they work. While there are sometimes differences in the degrees of participation that the two sides would prefer in the future, variations in the preferences as between different aspects of technical change, differences across the 12 Member States, there is nevertheless **a clear trend towards improved co-operation between the two sides and higher degrees of participation in the future.**

All our findings suggest that employee representatives — many of whom were trade unionists — are unreservedly enthusiastic about technological change.

Our results should be seen in the light of a number of political initiatives that are being taken to promote participation in European enterprises as part of the Social Dimension of the internal market in 1992. One important (and often under-rated) strand of European social policy is the Val Duchesse Social Dialogue. Our results were assessed against the various Joint Opinions which have emanated from this process. It is particularly significant that a recent understanding between UNICE and the ETUC has enabled Commission proposals on the Social Action programme to implement the Social Charter to be considered within the Val Duchesse process to see whether or not both sides can produce an agreement between themselves within a nine-month period before the Commission needs to issue a Directive. Our results can act as a benchmark in their deliberations.

When we compare the levels of participation against each other across this range of issues, we find that they are highest in health and safety and relatively high in work organisation and in training. They are lowest in product and service quality and in investment strategy.

The picture that emerges from a country comparison of participation levels in the aspects of technological change that have been explored in the survey show a distinct **North-South** divide among the Member States in the Community. Northern European countries such as **Denmark, Germany,** the **Netherlands, Belgium** the **United Kingdom** and **Ireland** all have higher levels of participation than those in the South. It is clear that participation is underdeveloped in the southern Member States — **Portugal, Spain, Italy, France, Greece** (with the inclusion of **Luxembourg** in this group).

Many mistakenly believe that participation can be created simply by introducing legislation or Community-wide Directives. However, while legislation **may** be the answer in some cases, it does not necessarily follow that the introduction of such legislation can **guarantee** that participation will occur in practice. There is ample evidence from our survey — particularly on health and safety — to suggest that the mere existence of legislation is no guarantor of participation. All this is to emphasise that **on its own** legislation is not necessarily the answer if we are to see improved levels of participation generally; it is merely **one** of a number of factors that are important. Participation depends on the willingness of both sides of industry in order to make it work. However, this is not to say that legislation supporting participation is always unnecessary; in some cases it it will be essential.

Most of our explanatory factors are likely to be favourable insofar as the further development of participation is concerned, both in northern and southern Europe. Given all the evidence from our survey results which show not only a strong trend towards greater degrees of participation in the future in new information technology but also a high degree of consensus on many issues, we can expect that as the Community moves towards greater economic and political union, the benefits to both sides of industry that are so clearly evident from their experience of participation in technological change will considerably improve relations between management and employees throughout Europe.

Chapter One

Introduction

1.1 Introduction

This report presents the results of an attitudinal survey on the impact of new information technology in enterprises across the 12 Member States of the European Community. It outlines the experience and opinions of nearly 7,300 managers and employee representatives from the countries of the Community. The survey covered five important industrial sectors; mechanical engineering and electronics in the manufacturing sector, and banking, finance and retailing in the services sector. The report concentrates on examining in some detail the impact that new information technology has had on jobs in European companies and the kind of participation that accompanies new information technology when it is introduced. This introduction gives an overview of the major research questions addressed in the survey.

The study is essentially comparative and the core questioning in the survey produced much valuable information about the perceptions of the respondents on the impact of new information technology on employment levels, particular occupations, employment flows within enterprises, the problems that new information technology posed for particular occupational categories, and the role of participation in dealing with these problems. A central focus of the study concerned the role of participation in technological change; in particular, the survey covered participation in training, health and safety aspects of new information technology, work organisation, investment, product quality and the benefits of participation to both managers and employee representatives.

This report is issued at a time when the Member States of the Community are moving towards the setting up of the Single Market. The Social Partners have emphasised the importance of creating a "Social Dimension" as part of this process to ensure that employees and citizens of the Community are given adequate protection from any adverse effects of this process that might arise. The Social Dimension is of major relevance to this survey and provides a benchmark against which many of the survey results can be measured. The European political context is discussed extensively in Chapter 3.

1.2 Major research questions of the survey

The introduction of new information technology in European enterprises during the 1980s has had profound implications both for employees and the way such enterprises are managed. Participation in technological change has been accorded a high priority by the Social Partners and thus the survey gives an unprecedented opportunity not only to evaluate the effects of these changes stemming from new technology, but also to give some detailed information on the degree of participation by employee representatives in such technological change. The survey results can also be evaluated in the light of the various political proposals that are currently coming into being.

In this context, the **European Foundation for the Improvement of Living and Working Conditions** in Dublin carried out an attitudinal survey in all member states of the European Community in 1987 and 1988. The survey was designed to measure the opinions of both managers and workforce representatives in selected industries about their perception and experience of participation in the process of introducing new information technology in their enterprises. The first wave of the survey focused on five EC countries: Germany, Denmark, France, Italy and the UK. This study[1] was published in 1989. The second wave of the survey[2] was broadened to include all 12 EC countries; this part of the survey consisted of two parts, a Euro-level analysis and a country comparative part. The Euro-level analysis looked at detailed aspects of the involvement of employee representatives in both the planning and implementation stages of the introduction of new information technology. The country comparative part consisted of a comparative evaluation of the levels of involvement of employee representatives in technological change.

This Report seeks to build on the findings of the two previous reports by examining in much greater detail the kind of participation that takes place within each Member State. This study is essentially comparative and extends the survey analysis in several ways. The **main topics** approached in this survey are the employment impact of new information technology, and the extent and the impact of participation on a number of issues (employment impact, training, health and safety, work organisation, investment, product quality) in each Member State at the time the interviewing was carried out. It further aims at finding out the potential for increased participation in these issues by employee representatives in the future on the basis of past experience. It also examines the benefits of participation to managers and employee representatives. In the survey, equal weight was given to the responses of both sides.

Chapter 2 of the Report gives a detailed explanation of the methodology employed in carrying out the survey. **Chapter 3** outlines the European political context within which the survey results should be seen. An important feature of the study is the use of an explanatory framework of country-specific factors as a means of explaining the differences in participation levels between particular Member States across the Community; this explanatory framework is outlined in detail in **Chapter 4**.

The following major research questions are covered in the survey:

1.2.1 The employment impact of new information technology (Chapter 5)

What movements of personnel took place in European enterprises as a result of the introduction of new information technology?

Did managers and employee representatives agree on the nature of these movements of personnel?

How are these movements of personnel in firms broken down at Member State level?

Is there evidence of job losses or job gains at both European and Member State level from the survey results?

Which particular occupational groups were affected by the introduction of new information technology, and to what degree?

Was there much concern shown by particular occupational groups about the perceived adverse effects of new information technology? Was participation a valuable process in alleviating these concerns? What differences of opinion between the two sides about these matters was shown in the survey results?

1.2.2 Participation in training (Chapter 6)

Training has been accorded a very high priority by both sides as a means of ensuring that employees have the right skills to face the competitive pressures in Europe in the 1990s. It is generally agreed that a common approach between the two sides is vital for training to succeed in this regard and thus participation levels in training matters are particularly important

What was the actual extent of workforce representative involvement in training in the European Community in 1987-1988? How is this broken down at Member State level?

How do the two parties perceive the intensity of past involvement? Do both sides have different or common views on the actual level of participation?

How does the actual intensity of involvement relate to the joint opinion reached at Val Duchesse? What is the proportion of companies which provide information and consultation in the process of technological innovation at both the European level and at Member State level?

What degree of participation was there in particular training issues (selection of participants, timing and duration, contents of training and type of training) at both European and Member State level?

What level of participation in training do both sides anticipate for the future, based on their past experience?

How can the different levels of participation at country level be explained by using an explanatory framework of country-specific factors?

1.2.3 Participation in health and safety (Chapter 7)

Another important topic which concerns all the Social Partners within the Community as we move towards the Single Market is that of Health and Safety. This is important because some of the Draft Directives which are being issued by the Commission as part of the Social Dimension are being drafted under the majority voting provisions of the Single European Act (Article 118A of the Treaty of Rome). How do our survey results relate to these Directives?

What degree of participation existed at both European and country level in health and safety matters?

Did the two sides agree on these degrees of participation?

What levels of participation do both sides anticipate for the future based on their past experience?

How can the different degrees of participation across countries be accounted for?

1.2.4 Participation in work organisation (Chapter 8)

Another part of the study focuses on participation in work organisation. It is well known that there are differences between management practice in EC countries with respect to how employees are managed and how work is organised in European firms. This is clearly an important concern of employees when new information technology is being introduced because of employee fears about deskilling, work intensity, changed work patterns etc. To what extent are employee representatives involved in decisions about work organisation?

What degree of participation existed at both European and country level in work organisation?

Did the two sides agree on these levels of participation?

What levels of participation in work organisation do both sides anticipate for the future based on their past experience?

How can the different degrees of participation across countries be accounted for?

1.2.5 Participation in investment in new information technology (Chapter 9)

Management has worries about employee influence in what they consider to be matters of managerial prerogative; investment in new information technology is one such issue in which managers believe that it is a matter solely of managerial prerogative. In this respect we consider data on participation in investment decisions concerned with new information technology.

What degree of participation existed at both European and country level in investment decisions?

Did the two sides agree on these levels of participation?

What levels of participation in investment decisions do both sides anticipate for the future based on their past experience?

How can the different degrees of participation across countries be accounted for?

1.2.6 Participation in product quality (Chapter 10)

We also consider data which we have on a matter which can be seen as important to both sides — that of participation in product quality. In order for European firms to compete in world markets, the issue of quality is of paramount importance. This is particularly so given the influence of Japanese penetration of world markets and the adoption by many European firms — especially in Britain — of Japanese management practices concerned with quality.

What degree of participation existed at both European and country level in matters of product quality?

Did the two sides agree on these levels of participation?

What levels of participation in matters of product quality do both sides anticipate for the future based on their past experience?

How can the different degrees of participation across countries be accounted for?

1.2.7 The benefits of participation to both sides (Chapter 11)

Finally, we examine the benefits of participation in a number of respects: the attention paid by managers to the concerns of employees and how both sides perceive it; the utilisation of the skills and knowledge of employees in implementing technological change in European firms; the time needed for decision-making by management when they are

INTRODUCTION

introducing new information technology in their enterprises; whether participation by employee representatives in such decision-making adversely affects the quality of such decisions and whether such decisions are made without being unnecessarily held up as a result of the participation process.

The survey findings throughout the Report will be assessed in the light of relevant Draft Directives, any relevant Joint Opinions at either European or sector level emanating from the Val Duchesse Social Dialogue and the views of both sides of industry within Europe. An important feature of the study is that the explanatory framework of country-specific factors will be used as a means of explaining the differences between particular countries in the Community.

The Report will conclude by setting out the implications of the survey findings for the Social Partners.

NOTES

1. *New Information Technology and Participation in Europe: the potential for social dialogue,* European Foundation for the Improvement in Living and Working Conditions, Office for Official Publications of the European Communities, Luxembourg, 1989.

2. *Roads to Participation in the European Community: increasing prospects of employee representative involvement in technological change,* European Foundation for the Improvement of Living and Working Conditions, Office for Official Publications of the European Communities, Luxembourg, 1991.

Chapter Two

Survey Methodology

The survey was designed to measure opinion of managers and employee representatives who were personally involved in the introduction of new information technology which had a significant impact in the workplace. The methodology was agreed upon by a group of experts including representatives of unions and employers at a European level.

In the analysis of technology introduction and its results equal weight was given to the information provided by, and opinions of, both managers and the employee representatives. This was achieved by ensuring that pairs of managers and employee representatives of the same company or establishment were interviewed. The survey sample was restricted to larger establishments (at least 50 employees) which had some kind of formal employee representation and which involved these representatives in the process of technology introduction. The response rate was around 50% and, as is common in surveys of this kind, there was inevitably a degree of self-selection inherent in those companies which took part in the survey.

1. Countries and Sectors

The study covers all twelve Member States of the European community. Within these countries, all industries and services introducing new technologies were eligible for the study. For practical and methodological reasons, we had to limit the coverage of the survey:

> Telecommunications was left out for practical reasons. This industry is interesting in terms of new information technology, but as it is highly monopolised and state-controlled in most countries, a survey would have posed unsurmountable practical difficulties.

> Car manufacturing as a major user of robotics and modern production control systems had to be excluded as it is not found in all European countries. Because of the lack of comparability, this major industry had to be left out of the survey.

The study concentrates on mechanical engineering and electronics in general to represent the industrial sector and on banking and insurance as well as retailing to cover the service sector. With these selections the main areas of new information technology application are represented: mechanical engineering covers a large and extremely differentiated area of activities which is established in all countries.

The mechanical engineering industry often uses electronics as a technical base, and in terms of practical applications, it is often difficult to separate both these forms of production from each other. Electronics is also considered to be a growth industry.

In the service sector, banking and insurance were chosen for two reasons: they were the first areas in the service sector to be heavily influenced by main-frame computerisation and which have experienced a second wave of new technology introduction with a new generation of more flexible computer technology. At the same time, banking and insurance also cover the vast area of work processing applications.

Retailing is the second area of the service sector that was included in the study. It potentially shares the word processing applications with banking and insurance, but it is also of particular interest in terms of new developments such as store control, sales analysis or purchase order decisions which have been made possible by the new information technology.

2. Survey Procedure

The point of entry into the company was the personnel manager or his/her functional equivalent. A short telephone interview was conducted with this person to establish the company's eligibility in terms of sector, size, and existence of some form of employee representation. If the company satisfied our criteria for inclusion in the survey, the personnel manager was requested to name managers and employee representatives who were personally involved in the introduction process.

The potential respondents were subsequently contacted by letter or by telephone to arrange personal interviews. Checks were made to ensure that employee representatives had played a role in the introduction process.

To reflect the national situation, there are minor variations in each country as to survey organisation.[1]

The survey procedure can be summarised as follows:

> a telephone interview with the personnel manager to establish the eligibility of the company for inclusion in the survey and to gain cooperation;

> an introductory letter sent to the potential respondents;

> personal interviews conducted with pairs of managers and employee representatives, preferably with two managers and two representatives of each company.

By this procedure, a total of 7,326 interviews — with 3,663 managers and the same number of employee representatives — were carried out in the 12 countries of the European community. The interviews were carried out in two waves: between 11 February and 22 May 1987 in Denmark, Germany, United Kingdom, France and Italy; in Belgium, Greece, Ireland, Luxembourg, Netherlands, Portugal and Spain between 5 April and 7 October 1988. Altogether, interviewing took place in 2,807 European companies which means that — as a statistical average — 1.3 managers and 1.3 employee representatives were interviewed in each company. This is the **largest** survey to be conducted on the topic of employee representatives' involvement in technology introduction in Europe.

3. Limitations of the study

On the whole, there was a very positive response to the survey by both managers and employee representations. However, two limitations of the study must be pointed out:

> Firstly, Ireland poses a special problem: very few establishments met the size or other criteria set for the survey. In spite of lowering of the standards of eligibility of companies, the number of interviews was particularly low in this country, the reason simply being the lack of establishments which were appropriate for the survey.
>
> Secondly, the interpretation of the data has to consider some distortions which are due to the screening procedure of the companies and the respondents within the companies. According to these procedures, representative results in a strict sense are probably not achieved. The possible bias in the data can be worked out in three steps:
>
> > The social partners, the Commission and the state representatives in the Working Group of the European Foundation agreed on a screening procedure that limited the eligibility of companies for the survey to firms which had formalised workforce representation. Because of this, companies without formalised workforce representation are systematically excluded. This first step in screening means that there is the possibility of over-representing in the sample companies which had a more positive climate of industrial relations.

The point of entrance to the firm was the personnel manager. He or she decided whether or not interviewing could take place. In the knowledge that workforce representatives would also be interviewed, he or she might have made the decision to take part in the study as a means of assessing the general attitude of his or her counterpart as to the issue of participation. We might, therefore, expect firms with a positive climate of industrial relations to be over-represented in the sample, while conversely, companies with conflictual industrial relations might be under-represented. For the same reasons, it seems likely that companies with rather non-conflictual technology introduction, leading to positive results, are over-represented. According to the screening rules agreed upon, the personnel manager nominated a workforce representative to be interviewed. This procedure might have increased the chance for a biased selection of workforce representatives: personnel managers might have nominated counterparts who were closer to their point of view.

NOTES

1. Full details of the survey procedure can be found in *Roads to Participation in the European Community: increasing prospects of employee representative involvement in technological change,* European Foundation for the Improvement of Living and Working Conditions, Dublin, 1991

Chapter Three

The European Political Context

3.1 Introduction

Many of the issues which arise in this report have to be seen within the context of political changes which are currently taking place as the European Community moves towards the setting up of the Single Market in 1992. An important feature of the move towards the Single Market is what has been called the Social Dimension. The Social Dialogue (the Val Duchesse process) and the Community Charter of Fundamental Social Rights for Workers are particularly important in relation to the issues explored by this survey of participation in technological change in Europe.

3.2 The Social Dialogue

The Social Dialogue between the two sides of industry at European level (the European Trades Union Confederation, ETUC, on the one hand, and the Union of Industries of the European Community, UNICE, and the European Centre of Public Enterprises, CEEP, on the other), was reactivated by Jacques Delors at the beginning of 1985. This type of Dialogue is covered by the Single European Act, Article 118B which states that 'the Commission shall endeavour to develop the dialogue between management and labour at European level which could, if the two sides consider it desirable, lead to relations based on agreement'.

The Social Dialogue was re-activated to try to secure areas of consensus between European employers and European trade unions on a number of matters which are directly relevant to our survey. The principles of this social dialogue are:

— a progressive approach to agreed solutions rather than the imposition of pre-ordained models;

— the use of flexible instruments like Joint Opinions, framework agreements and joint declarations rather than rigid solutions like directives;

— the primary initiative should be left to the social partners to search for common ground, with the European Commission playing a role of facilitator rather than of initiator.

Recently, the Social Dialogue has been seen as the main means of implementing many of the measures in the Social Action programme to put into effect the Community Charter of Fundamental Social Rights for Workers. Directives will only be issued in circumstances where the Social Dialogue process is seen as inappropriate for introducing such measures.

The first conference on the Social Dialogue was held in **Val Duchesse** in November 1985 on the initiative of the Commission, and new avenues for discussion between the social partners were opened. Eventually, in March 1987, these discussions led to the expression of a **Joint Opinion** on information and consultation as to the introduction of new information technologies in firms. In this document the social partners "recognised the need to make use of the economic and social potential offered by technological innovation in order to enhance the competitiveness of European firms and strengthen economic growth thus creating one of the necessary conditions for better employment and, taking particular account of progress in the field of ergonomics, for improved working conditions...".

The Joint Opinion of Val Duchesse emphasises the importance of information and consultation practices in European enterprises and leaves room for manoeuvre for both sides. Some would argue that it has led to very limited results, and there is a need to make the Joint Opinions which emanate from the Val Duchesse process binding on the parties.

Others would argue that it can be regarded as a first, open concept for social change within European enterprises and it should be seen as one of the main achievements within the social field.

The Social Dialogue has led to several Joint Opinions. Among them, the double agreement of 6 March 1987 on the new technologies (training and motivation, information and consultation) provides guide-lines of major importance for the two sides of industry throughout the Community, even though there are more developed systems of worker information and consultation in some Member States (for example, Denmark, Germany and the Netherlands) than others. The Joint Opinions on Training and Motivation (Chapters 6 and 11) and Information and Consultation (Chapters 6, 7, 8, 10 and 11) are both relevant to issues which are explored in this survey report.

In 1990-1991, further Joint Opinions were concluded on the "Creation of a European Occupational and Geographical Mobility Area and Improving the Operation of the Labour Market in Europe", "New Technologies, Work Organisation and Adaptability of the Labour Market", "Education and Training" and "The Transition from School to Adult Working Life". The first of these assigns primary importance to training throughout the working life of employees in order to promote a workforce which is better trained, more motivated, more mobile and able to adapt to change and the new qualification requirements (Chapter 6). The second and third of these Joint Opinions re-emphasised the importance of effective information and consultation procedures in training (Chapter 6), and also stated that 'all workers in the firm, whatever the size of the firm for which they work, should be entitled to the same health and safety protection at the workplace (Chapter 7). The results of this survey will be assessed in relation to the text of these Joint Opinions. Recently, the Social Dialogue has been extended to sectoral level, notably in construction, the public sector and transport.

What are the important provisions of the Social Dialogue in relation to our survey results? The most relevant Joint Opinions to our survey results are those provided in the double agreement of March 1987 on "Training and Motivation" and "Information and Consultation". The Joint Opinion on training and motivation provides that "...the process of introducing new technologies would be economically more viable and socially more acceptable if accompanied, among other things, by effective training and greater motivation for both workers and managerial staff, factors which, in their view, constitute a genuine investment...". The text then goes on to state: "... Information and consultation of the work-force or, depending on national practice, of its representatives, on training programmes carried out by the undertaking, would help to increase employees' motivation by improving their understanding of the changes facing the firm...".

The Joint Opinion on information and consultation stressed "the need to make use of the economic and social potential offered by technological innovation in order to enhance the competitiveness of European firms and strengthen economic growth, thus creating one of the necessary conditions for better employment, and, taking particular account of progress in the field of ergonomics, for improved working conditions". Indeed, this Joint Opinion places much emphasis on the need to involve all employees and/or their representatives in technological change so that they are both informed and consulted. The text of the Joint Opinion also states that this "information and consultation must be timely".

Many of the results of our survey can therefore be assessed in relation to one or both of these Joint Opinions.

3.3 Proposals of the European Commission for worker participation

The European Commission has put forward several proposals over the years for worker participation on the grounds that democratisation of company management can not only be a factor promoting industrial peace but also one which improves the quality of management itself. Attempts were made to introduce new directives relating to participation, in particular the so-called **Fifth Directive** which intended to regulate the structure and the activities of public limited companies, including workers' representation. The **Vredeling Directive** aimed at providing information and consultation rights for employees in multinational enterprises.

However, these attempts to regulate and extend workforce participation in Europe by institutional and legal forms of intervention have not been successful, at least for the present. At the moment, the enaction of the Fifth Directive is proceeding very slowly; the Vredeling Directive was first diluted, and then frozen by the council until 1989 for numerous and complex reasons.

Another proposal emanated from the Commission in June 1988; this was the adoption of a memorandum on the creation of a statute for the European company, which was sent to the Council, the European Parliament and the two sides of industry for their opinion.[1] This proposal — which would be optional — would allow undertakings access to a form of limited liability company directly linked to European Community law. This proposal put forward three alternative formulae for workers participation in the management of the European company; the German system, under which the workers are represented on management bodies; the Franco-Italian system of a works council representing the employees and quite separate from the management bodies; and the Swedish system, under which each firm lays down the rules for worker participation under an agreement concluded with the workers.

However, the implementation of the European Company Statute is by no means certain. The draft Directive under which it was introduced has as its legal basis Article 100 of the Treaty of Rome; measures introduced under this Article require unanimous voting in the Council of Ministers. There was opposition to this measure both by the UK government and UNICE and despite the support of many of the Member States for its provisions, its progess remains in doubt. Hence, whilst the European Company Statute might provide a means against which some of our survey results could be compared, this proposal is still very much under debate; as such it is difficult to use it as a benchmark against which to assess our results.

The latest proposal concerned with participation within European companies was the draft Directive on European works councils on the procedures for the information and consultation of workers of European-scale undertakings; the proposals cover European-scale undertakings, or groups of undertakings, with at least 1,000 employees and at least two establishments in at least two Member States, each employing at least 100 workers. The works councils could be triggered by employees or their representatives, or by management. However, this draft Directive was issued under Article 100 of the Treaty of Rome and therefore requires unanimity in the Council of Ministers for its adoption. Moreover, the draft Directive encountered opposition from both UNICE, who described the proposals as being "too limited, taking no account of national legislation, employers' authority, or of economic necessities", and the UK government, who saw its proposals as "inflexible and prescriptive" (see 3.4).

The European Commission has also issued a number of Directives concerned with health and safety at the workplace. Among them is the Framework Directive adopted in June 1989 on the introduction of measures to encourage improvements in the health and safety of workers at the workplace. This Directive establishes the framework for the new European Community policy on workers' health and safety. Its scope is very wide, covering both private and public sectors and laying down a number of general obligations to be met by the employer. Apart from its health and safety provisions, the Directive also deals with the disclosure of information to workers, means of consultation, and provisions for participation and training in health and safety matters.

In addition, the Commission has proposed a number of other Directives concerning information and consultation — all of which have already been put into practice. These include the "Collective Redundancies" Directive on procedures to be applied in redundancy situations, which was adopted in 1975; and the "Transfer of Undertakings" Directive, which provided for employees' rights and advantages in mergers and take-overs, which was adopted in 1977. Finally, the Commission published a Recommendation on equity-sharing and financial participation by workers in enterprises on 12 July 1991 as part of its work programme for 1991 on the Social Charter (see below).

3.4 The Community Charter of Fundamental Social Rights for Workers

Perhaps the most widely known strand of the European Social Dimension is the Community Charter of Fundamental Social Rights for Workers, which is popularly known as the Social Charter. This document was adopted by eleven of the 12 Member States at the December 1989 European Summit by means of a "Solemn Declaration". The European Commission intended to introduce proposals in the social domain before the end of 1992 and to select a first string of measures as part of its Action Programme for 1990. The Charter stipulates that "information, consultation and participation for workers must be developed along appropriate lines, taking account of the practices in force in the various

Member States... Such information, consultation and participation must be implemented in due time, particularly in the following cases:

— When technological changes which, from the point of view of working conditions and work organisation, have major implications for the work force, are included into undertakings;

— in connection with restructuring operations in undertakings or in cases of mergers having an impact on the employment of workers;

— in cases of collective redundancy procedures;

— when trans-frontier workers in particular are affected by employment policies pursued by the undertaking where they are employed."

This social charter was accompanied by an **Action Programme** which spelt out in detail the planned activities of the Commission up to 1992. One new initiative of the action programme in the field of participation is a proposal of the Commission on a draft directive for informing and consulting the employees of enterprises with complex structures, in particular transnational undertakings: "The Commission, following consultation with the social partners will prepare a draft for a community instrument which in substance could follow the following principles:

— Establishment of equivalent systems of workers' representation in all European-scale enterprises.

— General and periodic information should be provided regarding the development of the enterprise as it affects the employment and the interest of workers.

— Information must be provided and consultations should take place before taking any decision liable to have serious consequences for the interests of employees, in particular, closures, transfers, curtailment of activities, substantial changes with regard to organisation, working practices, production methods, long term co-operation with other undertakings.

— The dominant associated undertakings should provide the information necessary to the employers to inform the employee representatives."

The action programme stresses the belief that practices of good information and consultation of workers and their representatives is an explicit potential objective of the Commission and of most members of the Council of Ministers.

The Single Act sets out two provisions which have a bearing on our survey results. The first of these is article 118A, which gives the Community the power to pass Directives on a qualified majority voting basis in the Council of Ministers in order to encourage "improvement, especially in the working environment, as regards the health and safety of workers". The second provision is article 118B, which introduces into the Treaty the principle of social dialogue and the prospect of collective agreements between social partners at a European level. However, there is a major exception to the qualified majority vote provisions; article 100A stipulates that unanimity is still required for any proposal "which bears directly, indirectly or partially on the rights and interests of employed persons".

Since the 1989 European summit, the Commission has issued a number of Directives which have major ramifications with respect to many of the areas covered by our survey; significantly, many of them have been drafted under article 118A, which requires only a qualified majority vote on the Council for them to take effect.

Pursuant to its Action Programme, the Commission adopted three draft Directives on the subject of "atypical" work — part-time, temporary and seasonal employment. One of these three Directives will require unanimous approval in the Council of Ministers, but the other two will not. These Directives all contain provisions to different degrees which require undertakings to inform and/or consult with employees and/or their representatives on a number of matters concerned with training (Chapter 6), health and safety (Chapter 7) and the quality, timeliness and usefulness of the participation process in both health and safety and in training.

Similarly, a Directive which has already been adopted at the Council on the minimum health and safety requirements for work with VDUs provides for consultation and participation of workers and/or their representatives in both health and safety matters and in training.

The agreement between 11 of the Member States as part of the Maastricht treaty in December 1991 to extend qualified majority voting to a wide range of social and employment matters as part of the Social Protocol has given added importance to our survey results.

The Commission has already indicated that it intends to introduce many other proposals before 1992 as part of its Social Action programme. Thus, our survey results will be particularly useful in providing a detailed indication of the situation throughout the European Community with regard to a number of matters covered either by Directives which are currently being adopted or which are likely to be issued in the near future.

3.5 The views of the Social Partners

Whilst the Social Dimension of the 1992 Single Market is now receiving much attention in the Community as a whole, there is a wide divergence of views as to how the various strands of the social dimension should take effect. The political actors at the Community level (Commission, European Parliament, ETUC, UNICE) and the representatives of the two sides of industry on the company level (management, worker' representatives) have their own specific interests and policies on the social dimension of the Single Market. These divergent views are reflected in the differing positions taken on participation in technological change. Whilst there is a small degree of consent between all parties, there is also a great deal of disagreement.

The ETUC has defined its aims recently in 1985, 1988 and 1991. At its Fifth Congress in Milan in 1985, the ETUC stated that "effective negotiation of technological change often means influencing the change when it occurs and not merely reacting to the changes later. For the trade unions this means negotiating with management before the introduction has been planned. It means reappraising the knowledge and requirements of working parties in the production process and calling in outside experts to demystify the new technologies." At its Sixth Ordinary Congress in Stockholm in 1988, the ETUC defined its goals more precisely. Here, it asked for:

(a) "the right of employee representatives to be fully informed, consulted and also to negotiate on all important company matters before decisions are taken;

(b) equal participation by employee representatives in all company decisions of significance to the workforce;

(c) extension of decision making rights at all levels of decision making according to the organisation of companies ... The employee representatives at all plants must accordingly ... have the right to be informed and consulted on company planning, to negotiate and to represent their interest jointly at the European level." These policies were reaffirmed at its Seventh Ordinary Conference at Luxembourg in 1991.

At its 1991 Ordinary Conference the ETUC adopted a resolution which called for reinforcement of the Social Dialogue both by extending its content and by strengthening commitment to its role at three levels: European intersectoral level, with UNICE and CEEP; multinational company level; and cross-border regional level. The ETUC aims to promote the convergence of objectives through the co-ordination of bargaining strategies.

The ETUC position is strongly opposed by UNICE. UNICE insists on the right of company management to take final decisions. It further rejects any form of statutory regulation at the European level:

> "Employers do not oppose practices such as co-determination, let alone information and consultation. They strongly oppose the imposition of such practices on all Member States by means of EC level legislation. Such matters must be worked out by the parties most directly concerned, according to local conditions, traditions and legislation. In this way, the solution arrived at will be better, and the parties will be more deeply committed to making them work."

The bodies of the European Community, the **European Commission, Parliament** and **Council** pursue a policy which tries to reconcile the opposing views of both the ETUC and UNICE.

3.6 Discussion

The realisation of a Community social policy which has the support of all the Social Partners is indispensable to the balanced running of the Community. The Solemn Declaration on a Community Charter of Fundamental Social Rights taken at the 1989 European Summit represents a political commitment on behalf of the Council of Ministers. So far, the Action Programme to support the Social Dimension of the 1992 Single Market is proceeding slowly. However, there are indications that Member States are willing to accelerate the programme in 1991-2.

Finally, the European Foundation's survey data on participation in technological change is the only major survey that exists which gives valuable information on the levels and detail of participation in practice on a number of issues which are very much at the heart of the political changes that are currently taking place in the European Community. Our results can act as a useful benchmark in assessing current progress in the Action Programme and in pointing to areas which need attention in the move towards greater involvement by employees in the European organisations in which they work. Our results should be seen very much in this light.

In Chapter 4 we set out an explanatory framework which is a useful way of assessing the differing participatory practices in a number of areas (training, health and safety, work organisation, investment, and product quality) that exist across all 12 Member States as indicated in our survey results.

NOTES

1. The Commission proposed three alternative formulae for workers' participation in the management of a European company: a German system, in which workers are represented in managing bodies; a Franco-Italian system of enterprise committees where workers' representatives come together, separately from managing bodies; and a Swedish system where each company defines co-management rules through an agreement negotiated with workers' representatives. However, companies would not be enabled to opt for provisions circumventing a stricter national legislation already applying.

Chapter Four

Developing an Explanatory Framework Accounting for Inter-Country Differences in Participation in Technological Change at Country Level

4.1 Introduction

In this chapter of the Report we set out an explanatory framework which is designed to explain the different levels of participation that existed throughout the 12 Member States at the time the interviewing took place. This explanatory framework consists of a number of **factors** which assist or impede the possibilities of employee involvement in the introduction of new information technology and which are used to explain the survey data. However, it is important to note that whilst the survey data can be interpreted in the light of these factors, it does not always contain enough detailed information to relate the country differences to every one of these factors in every case; nor can the factors themselves be tested empirically.

There are a number of country-specific factors which shape the opportunities for employee involvement in the introduction of new information technology. It is important to mention at the start that some of these factors will not only be country-specific but will also apply at sectoral level and company level in the countries themselves. The factors will also be interrelated to some degree, and strong forms of particular factors might well have a dominating impact on the outcome of the degree of employee participation in technological change that emerges. These factors will not only shape the opportunities for participation but they will also affect the subject matter and timing of involvement. There are five principal factors which can help or hinder the opportunities employee representatives have for participation; they will not, however, determine the extent to which employee representatives can influence the effects of technological changes in the organisation. It is one thing to put discussion about technological change on to the formal agenda and quite another to translate these participative opportunities into a real ability to influence the way new information technology can affect working life.

4.2 Five principal factors

The extensive surveys and case studies that have been carried out in several countries in the European Community during the 1980s suggest that the opportunities for employee involvement in decision-making about new information technology depend on the following five main variables:

(a) — management's dependence on the skills and co-operation of its workforce to achieve its objectives for introducing the new information technology;

(b) — management style and its attitude to participation;

(c) — the bargaining power of organised labour to force management to negotiate or consult with its representatives in the absence of any voluntary disposition on the part of management to do so;

(d) — regulations which lay down participation rights for employees or their representatives on a range of matters at enterprise level;

(e) — the degree of centralisation of the industrial relations system that exists in the country concerned;

Each of these country-specific factors will have a number of favourable and unfavourable conditions attached to them, depending on the country concerned (see **TABLE 1** p. 54). Not only will these favourable and unfavourable conditions exert a primary role in influencing the strength of each of the five variables and hence in turn affect the level of participation that one might expect in a particular country, but also will affect other dimensions of participation, such as those dealt with in this Report, i.e. participation in training, health and safety, work organisation, investment, product quality etc. For example, we would expect higher levels of participation in work organisation in those countries where the trade unions were in a strong bargaining position, where they placed a premium on achieving beneficial forms of work organisation, and where there is a network of government or union-financed bodies concerned with the work environment.

Similarly, in those countries where there are works councils or equivalent bodies which have extensive rights with respect to health and safety matters either as a result of collective bargaining or legislation, we would also expect higher degrees of participation. There are other examples that could be cited which emanate from our explanatory framework; in those countries where there is a tradition of co-operation between management and employees and where management is dependent on the problem-solving skills and co-operation of its workforce in order to make the best use of new information technology, we would expect greater levels of participation in training schemes, particularly during the implementation stage of the introduction of new information technology.

4.2.1 Management's reliance on its workforce to achieve its objectives

The objectives that management has for introducing new information technology will obviously vary from one organisation to another. In most cases management will have several goals in mind when introducing new information technology into their organisation. The emphasis between these is likely to vary according to the priorities and purposes of their organisation and the context in which it operates. These goals include increased competitiveness, involving employees in the aims and "culture" of the firm, greater control over the work process, reduced costs, enhancing product quality, improving customer service etc. Whatever management's goals, the nature of technological innovation is of particular importance.

For example, companies which concentrate their activities either on the sale or production of standardised goods or services, which operate within mature price competitive markets or where office technology is used to reduce costs by standardising the collection or processing of information will utilise process innovations. The effect of such technological change in cases such as this is likely to lead to a more rigid segmentation of work and a greater concentration of programme-related functions which are carried out by specialist departments. In such circumstances, there is little need to involve their employees or their representatives in the planning or implementation of technological change unless they are forced to do so by the threat of industrial disruption or a reluctance by the workforce to accept the changes.

Conversely, where management is concerned to enhance the competitiveness of their organisation by offering its customers improved levels of service and enhanced product quality, the new information technology introduced will serve to highlight product performance, quality control, design requirements, technical sophistication, more flexible scheduling and greater emphasis on customer requirements. Such flexible systems require more adaptable and broadly skilled employees who are adept at problem-solving and whose skills and co-operation are highly valued.

Thus, management's dependence on its labour force for realising its objectives is an important factor in influencing the propensity shown by management to pursue a participative approach to change. The less management relies on the skill and expertise of computer specialists and senior managers and the more it relies on the skills, co-operation and problem-solving capacity of lower-level personnel, the greater the prospects of participation in technological change. For example, the premium which has been increasingly placed on maximising the commitment of employees to the "culture" of the company for which they work in the **United Kingdom** throughout the 1980s by the adoption of methods such as team working, briefing groups, quality circles, employee reports, company presentations within devolved organic structures coupled with the growth in joint consultation committees has led to considerably more participation of an *informal* kind in British companies.[1]

It is important to qualify all this by making a number of observations about the form of participation, its content and its timing. Participation will usually be of an **informal** kind, largely involving discussions with individual workers about job-related issues. Management itself will determine the stage at which employees are involved — usually at the early part of the **implementation** stage. Moreover, participation will involve **consultation** rather than negotiation. The parties, agenda and timing of participation is thus of management's own choosing.

Given the rapid pace of technological change and the shortening of product life-cycles, management dependence on the skills and problem-solving abilities of its labour force is heightened. In countries of northern Europe e.g. **Denmark** and **Germany**, where there is a long-established tradition of emphasising Quality of Working Life (QWL) programmes, that co-operation is often more forthcoming.

The countries where management relies strongly on the skills and co-operation of its labour force thus creating a favourable climate conducive to participation are **Denmark** and **Germany**, and to a lesser extent, **Belgium**, the **Netherlands** and the **United Kingdom**. These are countries where technological innovation tends to be more advanced and which have industries which rely on high product or service quality as a competitive strategy. In such cases, greater attention is given to design requirements, product performance and technical sophistication[2] — although it would be unwise to over-generalise these differences.

4.2.2 Management style and attitude towards participation

The pre-existing style of management will also have an important bearing on whether or not technological change is introduced by participative means or by managerial fiat. Traditions of co-operative industrial relations are marked by consultation and participation; a history of mistrust and antagonism is associated with an unwillingness on the part of management to have its prerogative challenged when it introduces technological change.

Whilst management attitudes, values and ideology can sometimes change in response to outside influences such as competitive product market pressures and the nature of the technology being employed, such an adaptation of managerial practices to environmental pressures can be difficult. In companies where managerial aversion to employee participation exists it tends to permeate through the whole management structure. At each level in the management hierarchy, the approach that individual managers take to employee participation will reflect the way they themselves are controlled from above. In such circumstances, any participation that does occur will be complementary to management's right to manage and make decisions, and not a challenge to that power.

Unless there is a very strong tradition of co-operation with employees' organisations, participation will be limited to employees rather than employee representatives and to involvement in task-related matters rather than strategic matters. Indeed, management may offer involvement by employees in job-related issues specifically to avoid the involvement by trade union or employee representatives. If employee representatives are admitted to the decision-making process it is likely to be at the **implementation** stage when the major design questions have been decided. Moreover, it is unlikely that the participation of employee representatives will be extended to issues such as the pace of work, job design, manning etc. let alone major strategic issues such as technological choice and investment.

There are wide variations in management style throughout the European Community. These range from the co-operative management styles which are generally found in northern European countries, particularly Scandinavian countries such as **Denmark**, and also to a lesser extent in **Germany,** the **Netherlands** and **Belgium** to those in southern European countries such as **France, Spain, Portugal** and **Greece.** Clearly, this is no more than a generalisation, and there are, of course, great variations from company to company, region to region, and sector to sector in each country.

4.2.3 The bargaining power of organised labour

An obvious factor which determines the extent to which trade unions and employees' organisations can gain access to decisions over technological change and new systems of work organisation is their actual power to force management to consult and bargain with them. This power will in turn depend on a number of other factors, the most influential of which include membership density, the willingness on the part of employees to act collectively in defence of their interests and their capacity to inflict damage on their employer. In some countries, of course, some of the trade unions themselves as a matter of policy deliberately refrain from being involved with management in technological change and sometimes this means they refuse to bargain about such change as well. Examples of such lukewarm attitudes to participation can be found among some unions in **France** and **Italy.** Survey evidence from the

United Kingdom shows that trade unions are more likely to use what bargaining power they have when technological change is accompanied by major organisational change.

It is important to understand that the issues which are thrown up in the workplace as a result of technological change do not readily lend themselves to the formal agenda of collective bargaining. The complexity and differential impact of changes in technology will often make it difficult to forecast its precise effects on skills, job content and job security. Moreover, its impact will vary from one job to another. Some employees may see the danger of being deskilled or perceive that their job security is being threatened whilst others may see themselves benefiting in terms of wider job responsibility, better promotion prospects or improvements in pay. In circumstances, therefore, where a trade union is facing technological change where its members are affected differently, it is difficult to present a united approach to management in the bargaining process, although the negotiation of a procedural agreement can be useful to both sides.

The evidence from a number of studies suggests that the degree of overall influence exerted by organised labour will be affected by five inter-related factors; these include the depth and strength of a union's organisational arrangements within the enterprise, the technical knowledge and expertise of its membership, the resources that it is able to apply to the task of developing alternative technological options, the ability to develop detailed strategies to guide and assist its membership in the process of change and finally, the capacity to threaten or possibly use sanctions to induce management to compromise on its proposals.

The organisation of the labour movement in the country concerned is of crucial importance in determining the bargaining power of trade unions. In countries where the trade union movement is united and not split along religious or political lines e.g. **Denmark, Germany** and the **United Kingdom**, trade union bargaining power is likely to be stronger. The degree of trade union density is also important; where it is higher e.g. in **Denmark, Belgium** and the **United Kingdom** this will facilitate the greater bargaining power. Another important factor is the nature of the relations between the trade unions and Labour or Social Democratic political parties; the close working relations (to varying degrees) between the DGB and the SPD in **Germany**, the LO and the Social Democratic Party in **Denmark** and the TUC and the Labour Party in the **United Kingdom** are all examples of enhanced bargaining power. Even where the trade union movement is split along religious or political lines, closer co-operation between the major trade union federations in the interests of greater unity may help bargaining power; recent developments which have promoted greater unity between the rival trade union confederations in both **Spain** and **Italy** are examples.

British experience shows that the existence of multi-unionism at enterprise level can considerably limit a united response to management. Because each group is represented by a different trade union, inter-union relations can be soured as a result and it is difficult for unions to bury their inter-union and inter-occupational differences. Moreover, it may also enable employers to mobilise the support of certain occupational groups and their unions who see opportunities to profit from the adverse effects of new information technology on others. Few of the New Technology Agreements which were negotiated in the **United Kingdom** during the early 1980s had more than one union signature on them. In countries where labour movements have been largely successful in implementing the principle of 'one industry one union' — for example in **Germany** — conflicts about contradictory interests between occupational groups are not absent but they can be dealt with within a single representative structure.

Scandinavian experience illustrates that a key factor in the ability of unions to influence the direction of change brought about by new technology has been the educational support programmes for their membership at local level and access to research and development information independent of employers. This has involved a conscious decentralisation policy on the part of national union leaders to involve shop floor members in technological change by drawing on their knowledge and skills and giving them support when they need it.

There are two critical points in the process of technological decision-making where organised labour needs to be represented: at the company level where the key investment decisions are made and questions of equipment selection and design are resolved, and at the establishment or workplace level where those decisions are implemented. An effective organisational structure, therefore, is one which enables both the negotiation of broader strategic issues at the company level and more detailed and specific issues at a localised level. In multi-establishment companies decisions to introduce new information technology — particularly when it has a negative effect on earnings or manning levels — is likely to be centralised. Therefore, organised labour needs a structure which can support negotiations at the **planning** stage of the introduction of new information technology; such a structure is essential for success in negotiations. However, employees' organisations can also achieve important concessions from management by negotiation during the

implementation phase of the technology at the workplace. The limited attention paid by central decision planners to the labour aspects of new information technology can sometimes provide considerable scope for local variations in matters such as staffing levels, the content and design of jobs, skills training, arrangements covering manning levels and the extent to which companies can enforce an intensification of work effort.

Denmark would appear to have the most potent bargaining power of organised labour of all the countries in the European Community — closely followed by **Belgium, Germany**, the **Netherlands**, **Italy** and the **United Kingdom**. Conversely, the bargaining power of organised labour is weakest in **Greece** (although the banking unions in **Greece** are particularly strong) and **Portugal**.

4.2.4 Legal Regulation

An important factor which is likely to shape the opportunities for employees to participate in technological decision-making is where there are rights in existence which oblige employers to provide their employees with appropriate information and to consult and/or negotiate with them. These rights take various forms and vary according to the legal systems of the countries concerned. They may be defined in the constitution, specifically laid down by statute, incorporated in collective agreements which have legal effect, enshrined in the powers of works councils which are given legal backing, or may simply be based on jurisprudence or common law.

Three points need emphasising in relation to the underpinning of participation by regulation. First, regardless of the form that such regulation takes there is **no guarantee** that effective participation will necessarily result from it. Second, the rights of participation that exist may not necessarily be found in any legal code; in some countries — for example the **United Kingdom** — the exercise of information and consultation rights can sometimes be subsumed under conventional collective bargaining without any legislative back-up. Moreover, not only is there a great deal of diversity in the regulatory provisions for participation from one country to another, but also there may be differences between enterprises and industrial sectors within the same country. Thirdly, it is important to emphasise that it does not necessarily follow that one particular form of regulation is more effective than others; for example, participation is strongest in **Denmark** where the regulatory procedures have their origins not in legislation but in effective collective bargaining.

In the majority of member states of the European Community a clear distinction is made between rights associated with **collective bargaining** (which may or may not be underpinned by legislation) and rights concerned with information, consultation and participation in enterprise decision-making which are often exercised under the auspices of **separate bodies** which deal with matters other than the negotiation of terms and conditions of employment. These separate bodies usually exist at the level of the company or the establishment and they often provide for participation in technological decision-making. Examples of such separate bodies are *Betriebsrat* (**Germany**), *Comité mixte d'entreprise* (**Luxembourg**) and *Semarbijdsudvalg* (**Denmark**). It is important to stress that the opportunities for employees to be involved in participation in decisions about new information technology are shaped by **both** collective bargaining and any separate participatory body that might exist. Indeed, there is a close relationship between the two.

The most common form of legal regulation for employee participation in the European Community is the Statutory Works Council. Such bodies can be found in various forms in eight of the EC countries — **Belgium, Germany, France, Greece, Luxembourg, Netherlands, Portugal and Spain**.[3] However, not only is there a great deal of diversity in the powers that such bodies have in providing for different levels of participation, but also the establishment of such bodies is not always mandatory.[4] In some countries the establishment of such bodies have to be triggered by the workforce in the company concerned. Moreover, there are different thresholds in the size of the workforce which have to be reached before such bodies can be established. In addition to Statutory Works Councils, there are also different provisions between Member States for employee representation on Company Boards.[5] Finally, there are other forms of statutory approaches to participation in the form of employee-only bodies which have statutory powers conferred upon them for negotiation and to be privy to certain information disclosure; examples of such bodies are the *consiglio di fabbrica* or *consiglio dei delegati* in **Italy** or the *groupes d'expression directe* in **France**.

4.2.5 The degree of centralisation of the industrial relations system

Regulation on participation in new information technology is also closely related to the political traditions in the country concerned which have shaped the industrial relations system. Countries where negotiations are centralised above the shopfloor level, i.e. are conducted for entire companies, industries, or cover the whole labour market tend to have a greater degree of industrial democracy. For example, Scandinavian countries and **Germany** have built on a long history of participatory management schemes. These traditions have developed within a general industrial relations context which can be described as a "social partnership" between management and labour.

This partnership between management and labour involves co-operation between the two parties in pursuit of joint goals, such as economic growth, which are perceived as the common basis for successful pursuit of their common interests, where both sides may moderate their demands so as not to jeopardise their joint goals. The State will seek to actively co-ordinate this process by the devolution of some State authority to organised interest groups, especially for the purpose of governing their constituents. In order for this State devolution to be effective, the interest groups must be organised comprehensively, with an internal concentration of private government, and representational monopoly granted by the State as a *quid pro quo* for the moderation of demands. It is important not to over-stress the formal nature of this kind of social partnership; in practice some of its strongest forms occur in an emergent and informal way and do not necessarily require deliberate action on the part of any of the participating parties to create formal structures in which corporatism takes place. Social partnership works best when both employers and trade unions are organised **centrally** i.e. when they are able to exercise authority over their respective affiliated organisations.

Within the European Community, both **Denmark**, **Germany** and the **Netherlands** — and to a much lesser extent **Belgium** — fulfil these criteria and such traditions of co-operation between the two parties have clearly contributed significantly to the planned introduction of new information technology in these countries. The industrial relations practices which are found in such countries provide a structure and framework, as well as a stock of necessary goodwill and centralised bargaining experience on all sides.

Generally speaking, we would expect to observe higher levels of employee representative involvement, less resistance to, and more consensus on technological change than would be found in countries which had more decentralised industrial relations systems. The *more* the industrial relations system is organised on strong *central associations* which enables the two sides to bargain at all levels within a central framework, the *more effective* the rules of participation will be.

4.3 Conclusions

In the preceding section of this report we have identified **five** variables which would appear to play an important part in shaping the opportunities for employee involvement in the process of technological decision-making: **management's reliance on its workforce** to achieve its objectives for introducing the new information technology; **management style** and its attitude to participation; the **bargaining power** of organised labour to force management to negotiate or consult with its representatives in the absence of any voluntary disposition on the part of management to do so; regulatory provisions which lay down **participation rights** for employees or their representatives on a range of matters at enterprise level; and finally the degree of **centralisation of the industrial relations system** in the particular country (see **TABLE 1**). Whilst not all these criteria can be used as a basis against which to assess the survey data on individual countries, and are no more than generalisations, they are clearly important in the overall debate on participation in technological decision-making.

TABLE 1: Factors affecting employee participation in technological change

Variable	Favourable conditions	Unfavourable conditions
Technological Objectives	Performance enhancement and problem-solving skills important for success	Cost reduction with little dependence on employees
Management Style	Co-operative	Conflictual and closed
Bargaining Power	Highly unionised; facing common technological threat, and strategically located; technically knowledgeable and skilled membership; united and cohesive union organisation	Multi-unionism, low unionisation, facing uncertain or variable impact from technology; lack of research resources; inexperienced officials; unions divided along political or religious lines.
Legal Regulation	Strong forms of law and other regulations	'voluntaristic' 'market forces' or weak forms of legislation
Industrial Relations System	Centralised	Decentralised

TABLE 2: Country-specific factors and their influence on participation levels in particular Member States

Variable	Favourable	Neutral	Unfavourable
Technological dependence	Denmark Germany Netherlands Belgium	Luxembourg France U.K. Ireland	Italy Portugal Greece Spain
Management Style	Denmark Belgium Netherlands	Luxembourg Ireland U.K. Germany	France Italy Spain Portugal Greece
Bargaining power	Denmark Netherlands Germany Belgium	France Ireland Italy Spain U.K.	Portugal Greece Luxembourg
Regulation	Germany Netherlands Denmark Belgium	Spain France Luxembourg Portugal Greece	U.K. Italy Ireland
Industrial Relations System	Denmark Netherlands Belgium Germany	Ireland Spain Portugal	U.K. Luxembourg Italy Greece France

---

NOTES

1. K.Sisson (Ed.), *Personnel management in Britain,* Blackwell, 1989.
2. M.Porter, *The Competitive Advantage,* The Free Press, New York, 1985; P.Adler, 'Managing Flexible Automation', *California Management Review,* Spring 1988; R.Hyman and W.Streek (Eds.) *New Technology and Industrial Relations,* Blackwell, 1988, pp. 25-26; A.Sorge, G.Hartmann, M.Warner and I.Nicholas, *Micro-electronics and Manpower in Manufacturing,* Gower, 1983.
3. A comprehensive guide to legal regulation of participation in the European Community can be found in *Legal Regulation and the Practice of Employee Participation in the European Community,* European Foundation for the Improvement of Living and Working Conditions, Dublin, 1990.
4. Statutory Works Councils are mandatory in Belgium, France, Luxembourg and the Netherlands.
5. Employees have statutory representation on company boards in Denmark, Germany, France, Luxembourg and the Netherlands.

CHAPTER FIVE

The Employment Impact of New Information Technology

5.1 Introduction

At the end of the 1970s and the beginning of the 1980s when new information technology was first being introduced into the workplace, one of the key topics of discussion was the effect that new information technology would have on employment. It was generally argued that there would be substantial job losses associated with the widespread use of new information technology, although this view was partly countered by the argument that falls in employment would be even greater if new information technology was not adopted, since competitors who did use it would have substantial market advantages.

The fears about job losses as a result of the introduction of new information technology were exacerbated by the fact that it was being introduced at a time when unemployment was rising in all the countries of the European Community following the demise of Keynesian demand-management economic policies in the early 1970s. It was also feared that since the new information technology was being introduced into both the manufacturing and services sector, it would no longer be possible to rely on the services sector to provide job opportunities to offset those jobs which were lost in manufacturing; new information technology was seen as a "heartland" technology where no industrial sector or occupation could escape its effects.

Earlier fears about the impact of new information technology in eliminating jobs on a wide scale have not, in fact, materialised; at least not yet. This has been borne out by a number of surveys and case studies including the DE/ESRC/PSI/ACAS survey in the UK by Daniel.[1] The OECD, in a major Report by a group of experts which was published in 1988, pointed out that the services sector, despite the widespread application of computerisation throughout the 1980s, had experienced substantial job gains.[2] In a major report to the West German government on the labour market effects of new information technology which was carried out by eight major research institutes, there was little evidence of *widespread* job loss as a result of the introduction of new technology.[3] Although the results indicated that there were slight job losses, it was also found that the risk of unemployment was much lower in those companies which had undergone major technical innovation. Despite the concern about the perceived adverse impact of new information technology on employment levels by many trade unions in Europe at the beginning of the 1980s, such worries currently occupy a much lesser prominence on the trade unions' agenda than they did at the beginning of the 1980s.

Whilst the empirical studies of the 1980s suggest results that run counter to the earlier predictions of major job losses as a result of the introduction of new information technology, it is important to emphasise that there is also little evidence from these studies which suggests that new information technology *creates* jobs on a *widespread* scale. In fact, for a number of reasons, it is extremely difficult to assess the impact of new information technology on employment levels empirically.

Firstly, the survey was carried out in 1987-88 — a more favourable time for lower levels of unemployment throughout the EC than exists at present. Secondly, the survey was restricted to certain sectors; it could well be that different results might have emerged had the survey been carried out in other sectors. Thirdly, attitudinal surveys can sometimes be misleading in the sense that the respondents are *per se* more likely to give a favourable response because of the degree of self-selection in the response rate.

There are other factors which have to be taken into account. For example, increases in personnel in innovating firms might be offset by decreases in other companies which continue to utilise conventional technology; Or new information technology might decrease the size of the workforces in companies where it is applied. But at the same time, the creation of new companies might have been facilitated through new information technology — thus creating new jobs in other industrial sectors. Thus, a balanced assessment of the net employment effects can only be done on a nation-wide basis taking job loss and job-creation in all sectors of the economy into account. An analysis based on single companies will certainly provide valuable information on these companies, but in these companies alone; such information will not give us any indication about the indirect positive or negative repercussions throughout the whole economy. When evaluating the employment impact of new information technology on the basis of our survey, we have to keep these data limitations in mind.

5.2 Employment effects at the level of the firm at European level: separate effects

The attitudinal survey of all 12 Member States sought information on the effects of new information technology on employment at the level of the firm. The survey sought information on how companies managed the introduction of new information technology in relation to their employees. The first management representative and the first employee representative in each of the 2,807 companies throughout Europe were asked:

(a) whether any employees were laid off as a result of new information technology;

(b) whether any employees left voluntarily as a result of it;

(c) whether any employees were transferred to another job that was not affected by new information technology;

(d) and finally, whether any new employees were recruited to work with the new information technology.

These four types of employment effects comprise all the possible employment *changes* due to the introduction of new technology. Accordingly, we can assume that in companies where managers and employee representatives reported no compulsory redundancy, no voluntary leaving, no transfers elsewhere in the organisation, no recruitment of new employees, the introduction of new technology did not have any employment effects at all. In other words, when the respondents denied any of the four effects above, they indicated a static, unchanged employment situation, even under conditions of technological innovation.

Figure 1: The employment effects of technology introduction according to managers and employee representatives -multiple responses (percentage)

	Managers	Employee represent.
No effects	38	39
Job loss	6	7
Voluntary leaving	10	11
In-company movements	30	30
Additional personnel	40	46

Source: Survey in all EC Member States, 1987-1988; 3 848 Managers and 3 848 Employee Representatives.

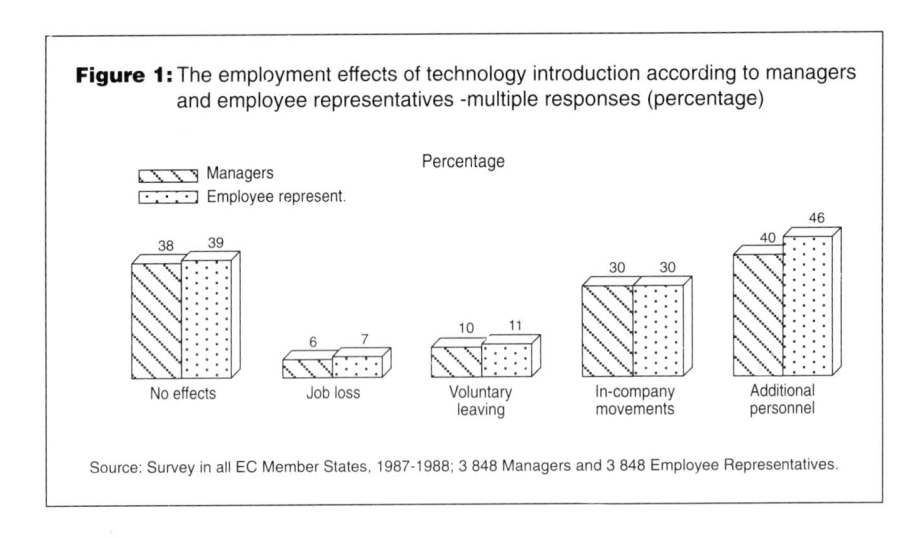

Figure 1 above sets out the responses to these questions at the overall European level. Perhaps the most striking feature of Figure 1 is the large measure of agreement between managers and employee representatives;

the difference between their responses is no more than 1% and thus we can be confident that we have obtained highly valid, factual information about employment effects.

Figure 1 indicates that the number of additional personnel recruited is by far the most important effect on employment of personnel, followed by "no effects" and employees being transferred to other jobs within the company. These responses vastly exceed the percentage of respondents who reported that "lay-offs" had occurred, although there was a somewhat higher response percentage of those who reported that employees had left the company voluntarily.

It might be argued that those respondents who reported that employees had left the company voluntarily were indicating that the severance of such employees from the company in fact represented a form of "hidden job loss". Even if this were the case, the percentage of respondents reporting that additional personnel were recruited vastly exceeds the combined percentages of those who reported "lay-offs" and "voluntary leaving". Our first indications then, produce a very positive portrayal of the employment flows in European enterprises when new information technology is introduced.

It is important to note that the percentages of all the responses in Figure 1 add up to more than 100% — in fact to 137%. The reason for this is obvious; respondents in answering this *multi-response* question were able to indicate that one, two or more effects on employment flows had taken place as a result of the introduction of new information technology into their companies. Such variability in employment flows is only to be expected in many companies. It is therefore necessary to attempt greater precision in our analysis about the effects of the introduction of new information technology on employment levels.

5.3 Analysing the employment effects with greater precision

We analysed our data to give greater precision in the analysis of the multi-response nature of the questions set out in 5.2 above. This gave us a frequency tabulation of the number of times various options were chosen by the respondents in combination with others. Given that there is a very high degree of congruence between both managers and employee representatives in their answers (the percentage difference between both sides even in regard to the combined employment effects never exceeded 2%), we analysed the responses of both sides together. The frequency tabulation indicated that four particular options were most commonly found:

(a) No employment changes at all (39%);

(b) Only new employees were recruited (25%);

(c) There were both new employees recruited and some job transfers (13%);

(d) Only job transfers within the company took place (10%).

The remaining options, which included the two forms of redundancy, accounted for no more than 13% of all optional responses.

The results are tabulated in Figure 2 below:

Figure 2: The combined employment effects of technology introduction

- 39% no effects
- 25% add. personnel only
- 13% add. pers. and moved
- 10% Pers. moved in comp.
- 13% others inc. job loss

Source: Survey in all EC Member States, 1987-1988; 3 848 Managers and 3 848 Employee Representatives.

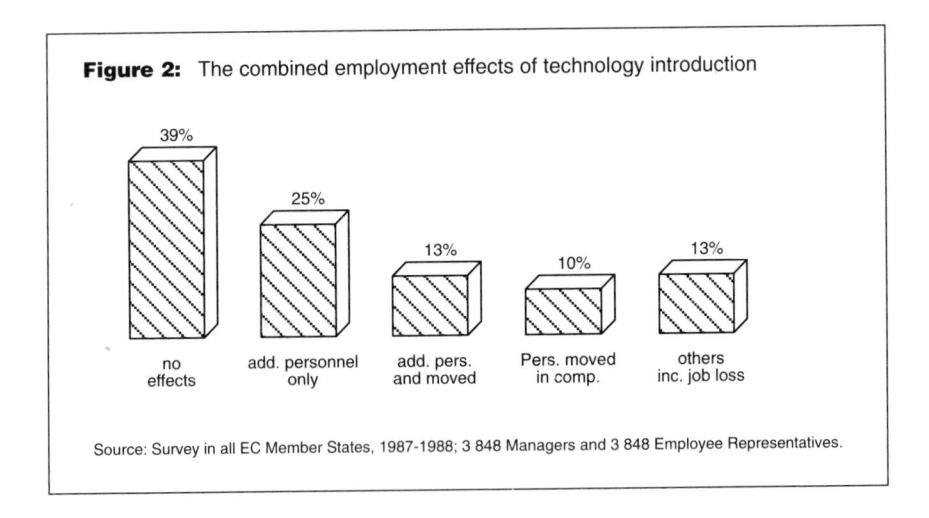

Figure 2 tells us that the most favoured option from respondents was that technology introduction did not necessitate any changes in the composition and utilisation of the workforce. Two out of five managers and employee representatives indicated that this was so. One in four respondents reported that the introduction of new information technology led to the recruitment of new employees; 13% stated that in their companies new employees were recruited while other employees were transferred to other jobs within the company, and one in ten respondents indicated that employees were solely transferred. A very striking result, however, is that the remaining possible combinations, including job loss and voluntary leaving amount to only 13% of all possible employment effects.

In sum, in about 60% of all cases there were employment effects due to the introduction of new information technology. Where there were changes, the hiring of additional personnel was the most prominent employment effect, followed by the combination of recruiting new employees and transferring others to other departments in the company. Job losses and voluntary leaving played a very minor role.

5.4 Combined employment effects at country level

In what ways are these employment flows reflected in individual Member States in the European Community? The similarity in the responses of managers and employee representatives were such that we were able to combine the responses of both sides into Figure 3, which portrays these flows in each Member State on a similar basis to the European level:

Figure 3: Combined employment effects due to technology introduction - Managers and employee representatives

Source: Survey in all EC Member States, 1987-1988; 4 321 Managers and 4 321 Employee Representatives.

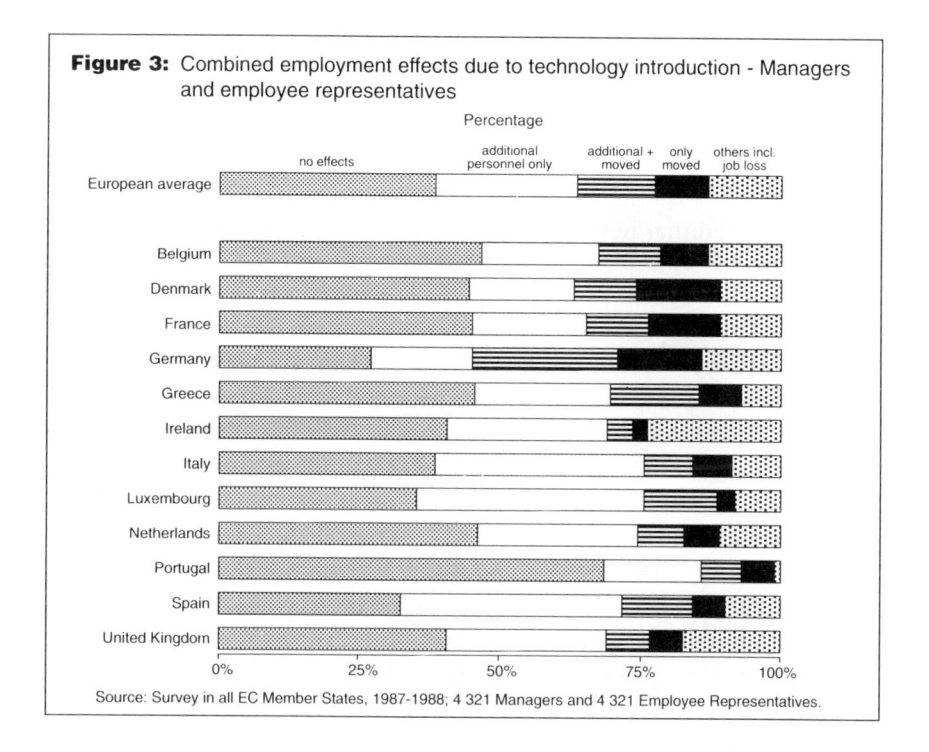

Figure 3 indicates that, in comparison with other Member States, employees in **Germany** are most affected by technological change. **Germany** not only has the lowest degree of "no change" (27%), but also a high percentage of respondents there report that new employees were being recruited and existing employees were being transferred to other jobs. Perhaps this is an indication of the advanced nature of technological development and investment in German industry during the 1980s. It might also reflect the importance of the internal labour market in that country.

In contrast, **Portugal** has the highest level of "no change" reported by its respondents as well as the lowest degree of redundancies. Firstly, it might well be a reflection of the low degree of technological innovation. Secondly, the Portugese system of industrial relations is extremely legalistic, and that might well explain the low level of redundancies.

The other conclusion which can be drawn from Figure 3 is that redundancies appear to be marginally higher in **Ireland** and in the **United Kingdom**. However, it is important to stress that when respondents indicated that some form of redundancy had taken place (the right hand column in Figure 3), such responses also possibly include other combinations (e.g. additional personnel recruited, transfers etc.). Finally, the percentages of respondents who reported that additional personnel had been recruited as a result of the introduction of new information technology was significantly higher in **Spain, Luxembourg** and **Italy**.

5.5 The quantitative employment effects of new information technology

Does this mean that new information technology has been instrumental in creating employment opportunities throughout the Community throughout the 1980s? Figures 1-3 have portrayed a very positive picture, with dismissals of personnel tending to be marginal while the recruitment of new employees stands out as the most prominent effect. But this result cannot be taken as proof that new information technology creates net employment gains. The data presented so far indicate only different *kinds* of employment effects. They give no indication as to the numbers of employees affected by these measures. For instance, respondents who mentioned that additional employees were recruited might have been referring to very small numbers and those who mentioned job losses might have meant very large numbers! If this were the case, we would have net job losses as a result of the introduction of new technology despite the overall positive figures so far presented. We can achieve a little more precision from our survey data in resolving this problem, but can by no means arrive at a definitive conclusion.

The questionnaire contained a question which asked respondents: "About how many employees altogether were affected by the introduction of this new information technology?". Whilst this question holds information on the numbers of employees affected, it does not relate these numbers to specific effects. In order to surmount this problem, we again analysed our data to give greater precision in the analysis of the multi-response nature of this question using the numbers of employees affected and relating them to the combination of employment effects which were reported by the respondents.

Because of the importance of this topic, we discuss below the procedure which we used in analysing the data. Up to now, we have used the information given by all respondents, i.e. all managers and all employee representatives. In order to arrive at the numbers of employees affected we have to use another method of data analysis and instead, analyse the data on a *company* basis in order to avoid the following problem: If we were to process the information given by both managers and employee representatives, we would in fact *double* the numbers of people affected if only one manager and employee representative were interviewed in a company. In the case of two respondents from each side being interviewed in one company, the actual figures would be *quadrupled*. Such an approach would clearly create an unrealistically high number of employees affected by new information technology. In order to surmount this problem, we have only used the information given by the first manager who was interviewed in each company; this information is used as the basis of our data.

At this point one might ask whether this information given by only the first manager can be taken as a valid, factual account of the quantitative employment effects. Our procedure pre-supposes an overall agreement between the first manager and the first employee representative in each company as to the number of people affected in the company. We applied diverse statistical tests to clarify this point. As we have already seen, there is an impressive consistency in the answers given by both sides in regard to the measures of employment changes. As can be seen from Figure 1, there is virtually no disagreement between the two sides in assessing the single kinds of employment effects. As far as the combined employment effects are concerned (of which Figure 2 gives the reduced version only), there are systematic differences between managers and employee representatives ($p < .001$), but the strength of the relationship is very weak (contingency coefficient = .08). This means in practice that when there were differences in the answers of both sides as to the occurrence of combined employment measures, these differences never exceeded 1.7%.

Further, the differences between both sides as to the reduced version of the effect combinations (see Figure 2) are due to chance ($p < .10$). Finally, we explored the differences between managers and employee representatives as to the numbers of people affected. The minimal differences in information are again statistically insignificant ($p < .16$), which means that there is an extremely high degree of consent between both sides as to the number of people affected by new information technology in the companies covered by the survey. Accordingly, we are on safe grounds to use the information given by the first manager as factual data for the respective company. We can also use the information given by the first employee representative and arrive at the same results.

According to the information given the absolute number of employees affected totalled 331,424 in 1,839 enterprises throughout the European Community. These numbers pertain to companies in which the introduction of new technology had some employment effects. Logically, these numbers cannot apply to companies where respondents stated that there were no employment effects. The frequency tabulation indicated that most of the employees affected were located in three particular combinations of options (a), (b) and (c) below chosen by the respondents in their answers to the question:

(a) only new employees were recruited (129,952 employees = 39%)

(b) employees were transferred elsewhere in the organisation and additional employees were recruited (85,627 employees = 26%);

(c) Only employees were transferred to other jobs within the organisation (56,283 employees = 17%);

(d) All other remaining combinations, including redundancy (59,562 employees 18%).

These results can be seen in Figure 4 below:

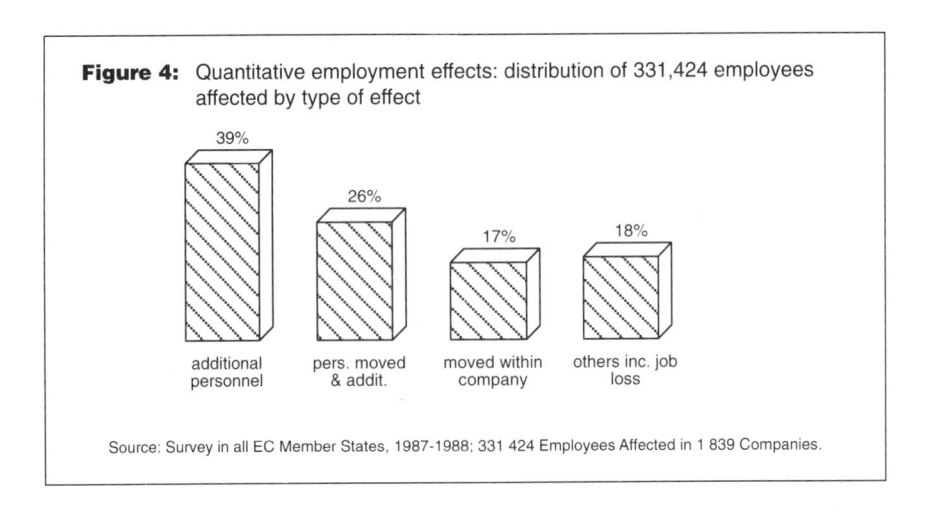

Figure 4: Quantitative employment effects: distribution of 331,424 employees affected by type of effect

- 39% additional personnel
- 26% pers. moved & addit.
- 17% moved within company
- 18% others inc. job loss

Source: Survey in all EC Member States, 1987-1988; 331 424 Employees Affected in 1 839 Companies.

The most striking conclusion that can be made from our results set out in Figure 4 is that, far from destroying jobs, the introduction of new information technology has proved to have had a very positive effect on job creation. Out of all 331,424 people in the European Community which were affected by the introduction of new information technology in the 1,893 companies in the survey, 39% were affected in the sense that they were newly recruited by the companies concerned. There are also elements of job creation embedded in the second combination (employees transferred and new employees recruited) without any precise indication of the exact numbers in the single measures. The category of the remaining combinations (including redundancy and voluntary leaving) account for 18% of the employees affected; these 18% account for just under 60,000 employees in total — an unknown number of which were victims of the introduction of new technology. Even if we were to count all of these 60,000 people as victims of new technology, the numbers concerned are rather small compared to the number of those who were newly recruited into the companies under study. Thus, it seems safe to conclude that at the time the interviews were conducted, the introduction of new information technology had very beneficial job creation effects.

Despite the very positive results at the European level, it would be wise to inject a cautionary note about these results. The first point to stress is that there is always a time lag in the impact of new information technology upon jobs; recent redundancies in the British banking industry resulting from heavy investment in new information technology during the 1980s are a good example of this. Daniel's survey[4] suggested that when new information technology is introduced into the workplace it was very rare for workers to be dismissed as a consequence of the introduction of such change. Generally, reductions in manning were managed through the redeployment in other parts of the establishment of people displaced by new machines. In our data, 17% of all people were affected in this way. Daniel then found indications that subsequently, in the longer term, the introduction of new technology did result in employment reductions after a time.

The second point to stress is that it would be unwise to attribute the employment changes emanating from our survey directly or only to the impact of technological change. The respondents to our questionnaire were prompted to attribute changes in employment flows to the introduction of new technology; some of these respondents might have indicated that the changes in employment flows were the result of new technology introduction when such changes had occurred for some other reason. It is notoriously difficult to isolate employment changes resulting from technological change from those which can be attributed to other factors, e.g. loss of demand for goods or services, reorganisations etc. The questionnaire on which our results are based did not contain any questions to managers and/or employee representatives about manpower reductions which were the result of these other factors.

Thirdly, perhaps the most important cautionary note relates to the selection of companies which were screened and which took part in our survey. All the companies which were interviewed in our survey had to meet certain criteria. Failure on any of these criteria rendered the company ineligible for inclusion. All companies must have introduced at least one new information technology from a specific list of technologies within the five years preceding the interviews in 1987-88; they must also have a formal system of employee representation and they must have involved their employee representative(s) in the introduction of new information technology. The use of such criteria would make it much more likely that many of the companies which took part in the survey were successful and profitable, had few problems in introducing new information technology, had a good industrial relations climate, and had little to conceal from interviewers.

Finally, when interpreting these positive results as a general indicator that new information technology creates more jobs than it destroys, we have to consider the possible indirect effects which cannot be assessed by our survey. Negative impacts resulting from new technology may be felt outside the sections and companies where it is directly applied, in companies which lose their competitive advantage by not applying new technology and which might have to dismiss their employees as a result.

Nevertheless, even taking into account the above limitations, the picture that emerges is a very positive one, at least in the sectors in which the survey took place (see chapter 2). It seems that the introduction of new information technology into enterprises does not lead to the elimination of jobs on the scale which was feared at the beginning of the 1980s. It might be that when the new information technology is fully operational in the longer-term that this apparently positive picture might pale somewhat, but that possibility is a matter of speculation.

5.6 Occupations affected by technological change

There have been very few studies so far which have sought to identify particular occupations which are most at risk as a result of the introduction of new information technology. In any case, such studies have to be treated with some caution because the occupations concerned will not necessarily disappear as such but might simply be changed in some way — either by a form of deskilling or by being subsumed into other tasks which were formerly carried out by other occupational categories. It is also important to distinguish between the short-term and long-term changes involved.

One important feature of the new information technology is that it affects the whole of the enterprise and not just particular parts of it. This is because of the importance of *information* in management decision-making. Unlike previous technologies, which only affected particular sectors and particular occupations, the new information technology is a "heartland" technology and it thus has pervasive effects throughout the organisation. In addition, case studies and surveys on the effects of new technology on employment[5] suggest that the more unskilled a particular occupation, whether in the office or on the shopfloor, the more likely such an occupation will be at risk.

We would expect, therefore, that our survey results would indicate that a large percentage of particular occupational groups would be affected by technological change.

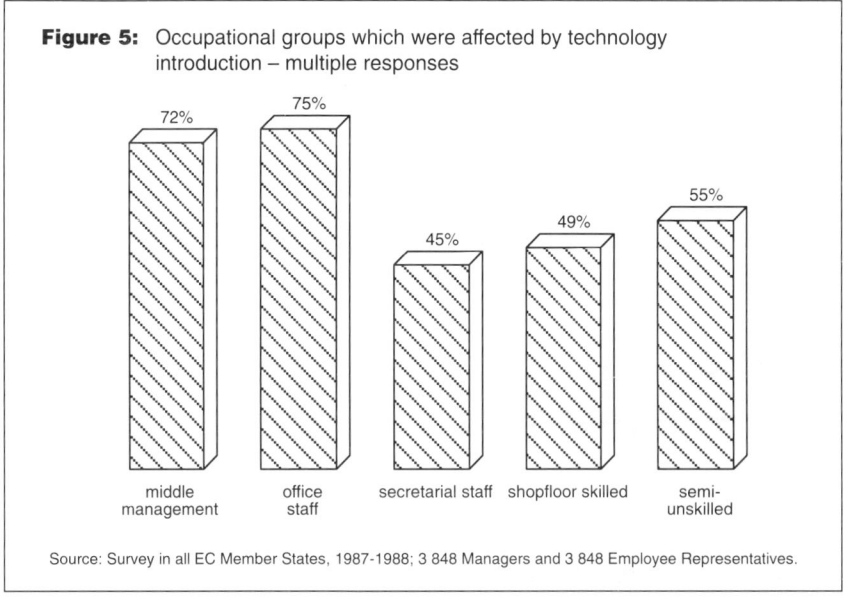

Figure 5: Occupational groups which were affected by technology introduction – multiple responses

- middle management: 72%
- office staff: 75%
- secretarial staff: 45%
- shopfloor skilled: 49%
- semi-unskilled: 55%

Source: Survey in all EC Member States, 1987-1988; 3 848 Managers and 3 848 Employee Representatives.

The questionnaire asked respondents: "some groups are sometimes said to feel the effect of technological changes more than other groups. Which of the following groups were directly affected at all by this new information technology?". As with the questions on the employment impact of technological change, this was a *multi-response* question. In Figure 5 above we illustrate the frequency that both managers and employee representatives mentioned that particular groups were affected at the European level.

Figure 5 above shows that the introduction of new information technology did have a pervasive effect on European enterprises with large percentages of particular occupational categories affected. The Figure indicates that middle management and office staff are most affected, with secretarial staff less so; just over half of all shopfloor skilled and shopfloor semi/non-skilled workers felt the effects of technological change.

Although we have combined the responses of both managers and employee representatives in Figure 5, there is little difference between them; the variations account for no more than 7% at the most.

5.7 The role of participation in ameliorating the effects of new information technology on particular groups — European level

The questionnaire was designed to find out whether these five occupational groups who were affected by the introduction of new information technology were worried about its effects on their jobs. Respondents were asked: "When this new information technology was introduced, did members of any of these groups express concerns or complaints (either directly or through their representatives)?". This would give us an idea of the extent to which new information technology was causing problems for particular occupational groups.

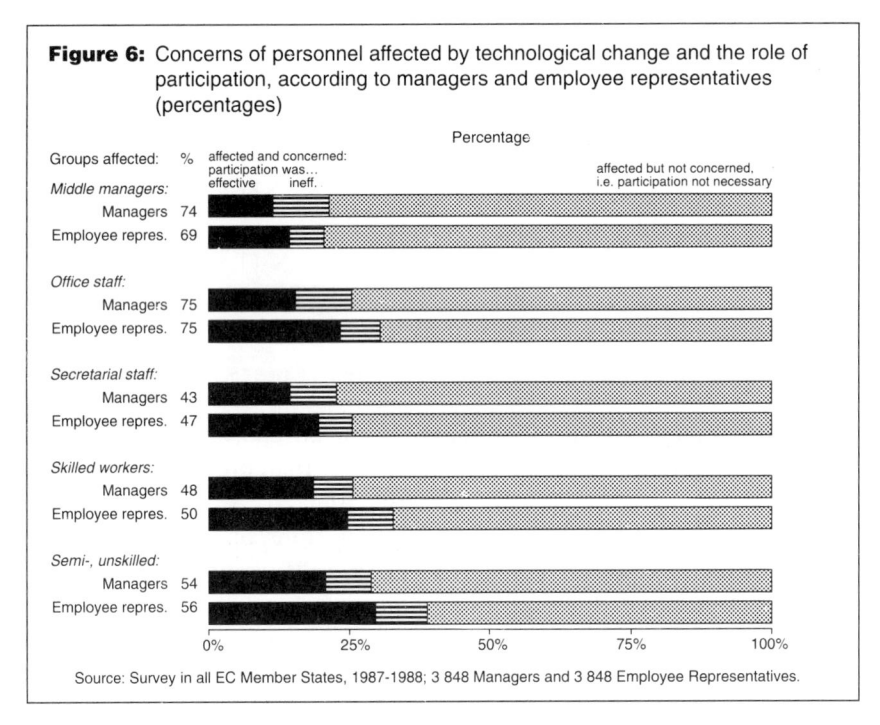

Figure 6: Concerns of personnel affected by technological change and the role of participation, according to managers and employee representatives (percentages)

Groups affected	% affected and concerned: participation was effective / ineff.	affected but not concerned, i.e. participation not necessary
Middle managers: Managers	74	
Employee repres.	69	
Office staff: Managers	75	
Employee repres.	75	
Secretarial staff: Managers	43	
Employee repres.	47	
Skilled workers: Managers	48	
Employee repres.	50	
Semi-, unskilled: Managers	54	
Employee repres.	56	

Source: Survey in all EC Member States, 1987-1988; 3 848 Managers and 3 848 Employee Representatives.

At the same time, we were also interested to find out if participatory machinery was effective in dealing with any complaints or concerns that arose. Hence the questionnaire asked: "Did the involvement of employee representatives in the introduction process help any of these groups to reduce any real or perceived adverse effects from it?".

The most striking feature of Figure 6 on p. 70 is that while large percentages of employees in the enterprises covered by our survey were directly affected by the introduction of new information technology, only a **small** percentage of those affected expressed concern about its impact. In general, it seems that all occupational categories adapted well to technological change — a finding in line with Daniel's survey in the U.K.[1] The percentage of those who were affected and who expressed concern about what they saw as the adverse effects of technological change varied from one occupation to another, with semi/unskilled workers expressing most worries. Figure 6 also shows that employee representatives reported that occupational groups were affected and concerned more so than their management counterparts. At the same time, employee representatives viewed participation as somewhat more effective in dealing with the adverse effects of new information technology than managers. Obviously, employee representatives were more aware of the negative repercussions of new technologies, and tended to evaluate their own role in managing such problems in a more positive light than their management counterparts.

We discuss the role of participation in dealing with those employees in the various occupational categories who were adversely affected by the impact of new information technology at the aggregate European level below:

5.7.1 Middle Managers

At the aggregate European level middle managers were highly affected by the introduction of new information technology; 74% of the management respondents and 69% of their employee representative counterparts indicated this. But at the same time, only 24% of the middle managers who were affected were said to have expressed concern about its effects on their jobs. However, whilst both sides were in complete agreement about the effects of new information technology, there are differences between them about the value of participation in resolving these concerns. Employee representatives were more optimistic than their managerial counterparts that participation was a valuable means of resolving these concerns.

It is not surprising that so many respondents reported that middle management had been affected by technological change. One of the features of economic development during the 1960s and 1970s in the Western industrialised nations was a general trend towards larger, more complex organisations. The problems of management organisation in these organisations increased as companies grew in size and diversification, with the need to employ more and more managerial, specialist, technical and administrative staff with the consequent elongation and complexity of management structure. However, new information technology spawned novel ways of dealing with the problems of management complexity, as organisations sought to become more flexible in their management organisation. Indeed, many commentators emphasised that new information technology would not only offer *more choice* in management structure, but also it could permit a general *contraction* and a radical *simplification* of such structures.[6] Whilst our survey results do show that middle managers have been affected on a widespread scale by the introduction of new technology, very few of them are worried about the impact of technological change.

5.7.2 Office Staff

According to both managers and employee representatives, office staff were affected slightly more than middle managers (75%); nevertheless, few of them were said to be overly concerned about this. Less than one third of all office staff affected voiced concern about their situation. For this group, participation helped a great deal in resolving their problems, and employee representatives were more supportive than managers about the effectiveness of participation in dealing with their problems.

We would expect a higher percentage of this occupational group to be worried about new information technology; technological change in the office environment has proceeded at a rapid pace throughout the 1980s. Office workers are less likely to be represented by trade unions and any participatory machinery is likely to be informal, individually focused and localised in character. Daniel's survey in the U.K.[4] found that office workers, contrary to the widespread view that they are inherently more favourably disposed to technological change, were no more favourably disposed to new information technology than their manual counterparts. We would expect, in the light of Daniel's findings, that office workers in enterprises where participatory machinery did exist (as was the case in this survey) would be more likely to express concern about the effects of new information technology. Whether our findings fall in line with this expectation cannot be assessed from our data, as our sample does not permit any comparison with companies without participative machinery.

5.7.3 Secretarial staff

Unlike middle managers and office staff, secretarial staff have generally been much less affected by technological change. According to managers, only 43% were affected while employee representatives see the situation of secretarial staff as somewhat more problematical (47%

affected). On the whole there is the somewhat surprising result that new technology affects the work of less than half of office staff. In addition, of those who were affected, only about a quarter expressed concern, the other three quarters obviously had no problems with the changed job situation. For the majority of the ones who were affected and expressed concern, participation turned out to be a helpful mechanism in problem solving — although again, employee representatives were more enthusiastic about the effectiveness of participation than managers.

5.7.4 Shopfloor skilled workers

Skilled shopfloor manual workers have also experienced a great deal of technological change during the 1980s. According to our data, every second workplace was affected by new information technology. Thus, the work situation of shopfloor skilled workers is very similar to that of secretarial staff. Like secretarial staff, shopfloor skilled workers seemed to have coped with technological change quite well. About two thirds of employees in this occupational group did not voice any complaints about new technology. However, unlike the situation of the other occupational groups discussed so far, managers and employee representatives are in greater disagreement about the numbers of shopfloor skilled workers who were affected and concerned: one in four managers stated that shopfloor skilled workers were affected and concerned, compared with one third of employee representatives. However, both sides are more or less in accord that participation mechanisms had mainly positive effects in smoothing out the problems that arose. All in all, we would have expected that shopfloor skilled workers would be slightly more concerned about the effects of new information technology on their jobs than they actually were. There was little evidence from our survey results to suggest that this occupational group overly feared any adverse effects of new information technology on their occupational situation.

5.7.5 Shopfloor semi-skilled and unskilled workers

Studies of the impact of new information technology on occupational groups have shown that the less skilled and least well educated employees are much more likely to be adversely affected by technological change. We would expect, therefore, that this occupational group would be more likely to express misgivings about the effect of new information technology on their jobs. Our survey suggests, that this is true to a limited degree only.

Semi-skilled and unskilled workers were somewhat more likely to be affected by new technologies (55%) than skilled workers (49%) and they complained about the effects of the introduction of new information technology more than any other occupational group. But in spite of these negative outcomes of technology introduction, we must not overlook the fact that 45% of semi-skilled and unskilled workers were said to have been unaffected by new technology and that almost two thirds of the ones affected did not voice concern over this fact. As to the size of the subgroup of workers which were affected and expressed concern, the information given by managers and employee representatives differs widely, with managers indicating a share about 30% compared with a figureof 40% by employee representatives. Both sides of industry are in accord about the effectiveness of participation as a means of resolving any problems arising from new information technology — employee representatives slightly more so than managers.

What kind of picture emerges from our survey at the European level about the perceived impact of new information technology on particular occupational groups and the virtues of participation in resolving any problems that arise? Perhaps the most important finding is that middle managers and office staff are distinctly more affected by new information technology than any other occupational group.

Secondly, there is a large majority of employees in all occupational categories who do not feel that new information technology poses any threats to their jobs although their jobs have been affected. Unskilled and semi-skilled manual workers are most concerned about technological change, but even here a majority of these employees show little concern about it. These results are most positive and it seems that the introduction of new information technology in European enterprises during the 1980s has enjoyed much support from employees. This is entirely consistent with the findings of Daniel's survey where it was shown that, far from opposing technological developments, workers generally supported technological change.

The third point to stress is that there appears to be some disagreement between the two sides both about the degree of concern which is felt by different occupational groups on technological change and the value of participation as a means of solving employees' problems. Managers generally tend to slightly underestimate the degree of adverse effects and are more sceptical about the usefulness of participation in resolving problems.

Despite these small differences between the two sides, it is clear that participation nevertheless enjoys widespread support from both sides as a means of dealing with the problems stemming from the introduction of new information technology; this was the case across all occupational groups.

5.8 The role of participation in ameliorating the effects of new information technology on particular groups — country comparisons

We were also interested to discover on a comparative basis the degree to which new information technology was perceived as a threat to particular occupational groups in different Member States of the Community. In doing so we compare the responses of managers and employee representatives. However, in some countries this means that we have to deal with small numbers with all the implications this poses for gaining meaningful results.

This is particularly evident in the case of **Ireland**. It is important to interpret the Irish results with a great deal of caution. The number of respondents interviewed there was very small compared to the size of the sample for other European Community countries so that, even allowing for the weighting procedure in our methodology, too much significance should not be attached to the results.

Nevertheless, the results for the various occupational categories at Member State level provide some interesting comparisons:

5.8.1 Middle managers

New information technology has clearly had a pervasive effect on the jobs of middle managers throughout the Community. Both sets of respondents are in close agreement on this point, and the large differences in opinion between the two sides in the cases of Ireland and Luxembourg can be accounted for by the small numbers of respondents involved.

If we ignore the results of **Ireland**[7] *and* **Luxembourg**, the pervasiveness of new information technology is particularly notable in the **Netherlands, U.K., Belgium** and **Denmark**; all these countries are above the European average of 74%. If we consider the responses of managers first, it is notable that in the **United Kingdom** and in **France** management respondents report that a larger than average percentage of middle managers felt threatened by new information technology; in **Portugal** hardly anyone was concerned about technological change. As far as the value of participation as a means of dealing with employees' problems is concerned, participation was viewed very positively by respondents in most countries — particularly so in **Greece**. On the other hand, a majority of management respondents in two countries — **France** and **Germany** — took a more negative view of participation.

Figure 7: Concerns of personnel affected by technological change and the role of participation, by countries (percentage)
— middle management, according to managers

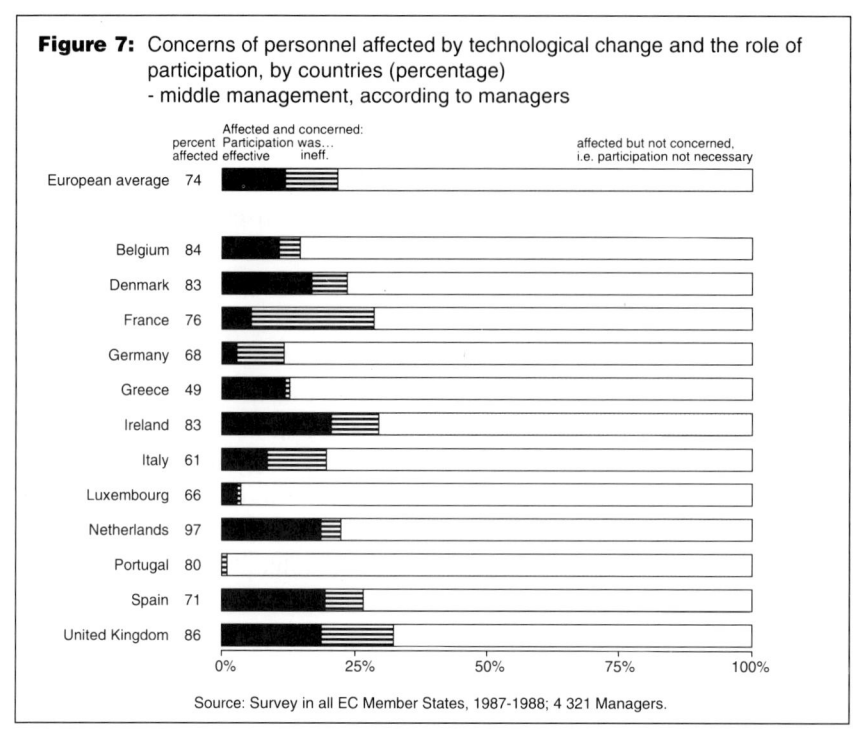

Country	percent affected
European average	74
Belgium	84
Denmark	83
France	76
Germany	68
Greece	49
Ireland	83
Italy	61
Luxembourg	66
Netherlands	97
Portugal	80
Spain	71
United Kingdom	86

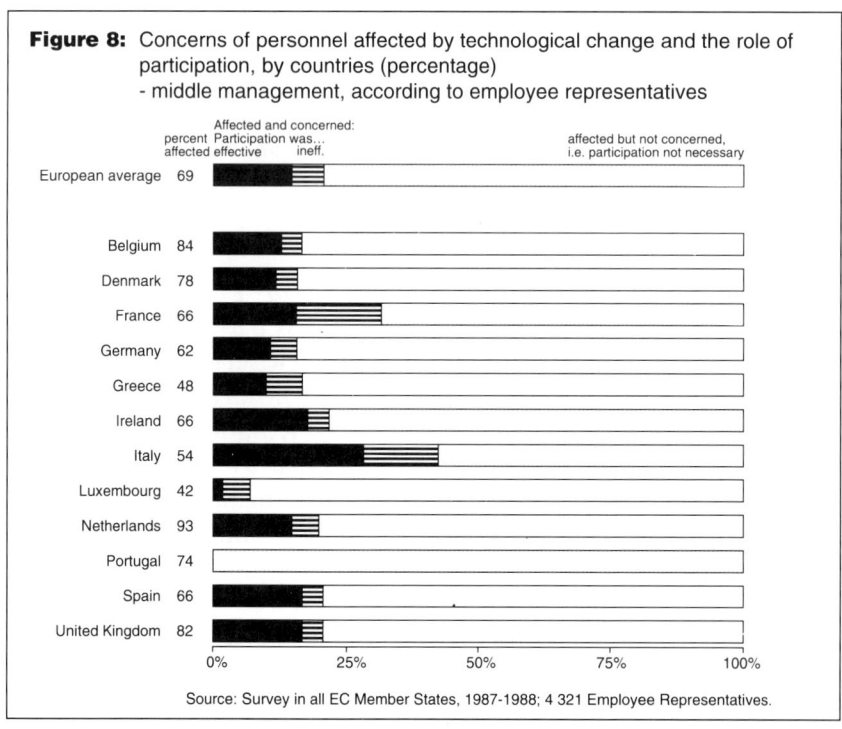

Affected and concerned: Participation was... effective / ineff.
affected but not concerned, i.e. participation not necessary

Source: Survey in all EC Member States, 1987-1988; 4 321 Managers.

Figure 8: Concerns of personnel affected by technological change and the role of participation, by countries (percentage)
— middle management, according to employee representatives

Country	percent affected
European average	69
Belgium	84
Denmark	78
France	66
Germany	62
Greece	48
Ireland	66
Italy	54
Luxembourg	42
Netherlands	93
Portugal	74
Spain	66
United Kingdom	82

Source: Survey in all EC Member States, 1987-1988; 4 321 Employee Representatives.

There was little difference between the views of employee representatives and management respondents in their views on the percentage of middle managers who were affected by technological change. Employee representatives in **Italy** believe that new information technology has caused concern for middle managers to a much greater extent than their management counterparts. As to the value of participation, employee representatives took a much more positive view of the value of participation in dealing with the adverse effects of new information technology than managers in nearly all Member States. This was particularly so in three countries — **Italy, France** and **Germany.**

5.8.2 Office Staff

Around three quarters of managers and employee representatives report that technological change has affected office staff — particularly so in the **Netherlands, Portugal, Germany** and the **U.K..** If we ignore the results of **Ireland** and **Luxembourg** because of the small number of cases, about three in every ten management respondents claim that office staff are concerned about technological change; **France** and the **U.K.** have the largest percentage of office staff who are worried about technological change. Employee representatives, in contrast, claim that more office staff were concerned about technological change.

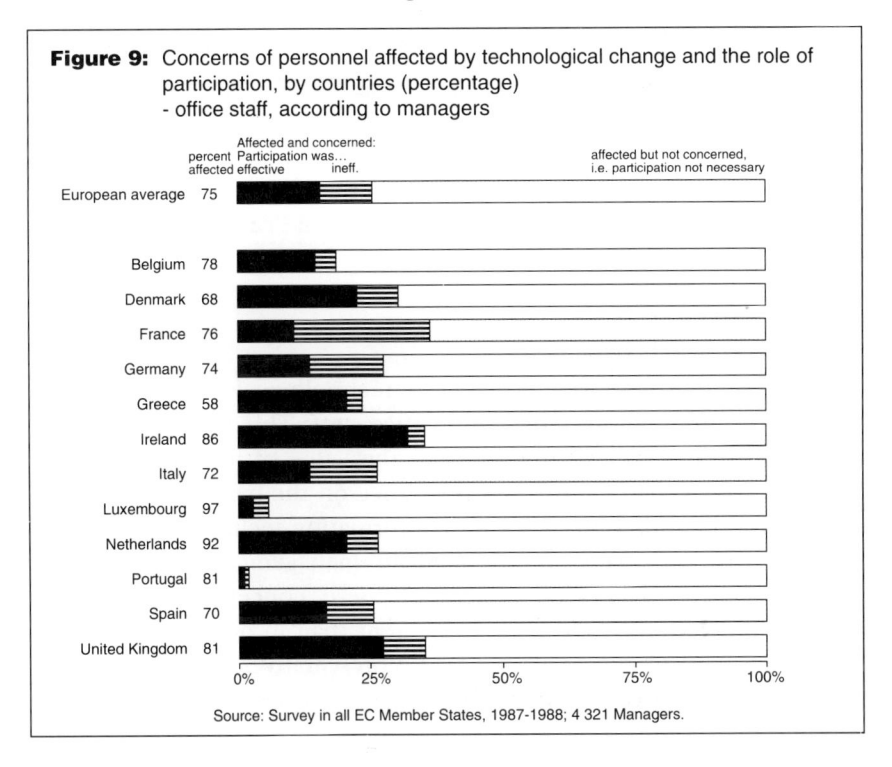

Figure 9: Concerns of personnel affected by technological change and the role of participation, by countries (percentage)
- office staff, according to managers

Country	Percent affected
European average	75
Belgium	78
Denmark	68
France	76
Germany	74
Greece	58
Ireland	86
Italy	72
Luxembourg	97
Netherlands	92
Portugal	81
Spain	70
United Kingdom	81

Source: Survey in all EC Member States, 1987-1988; 4 321 Managers.

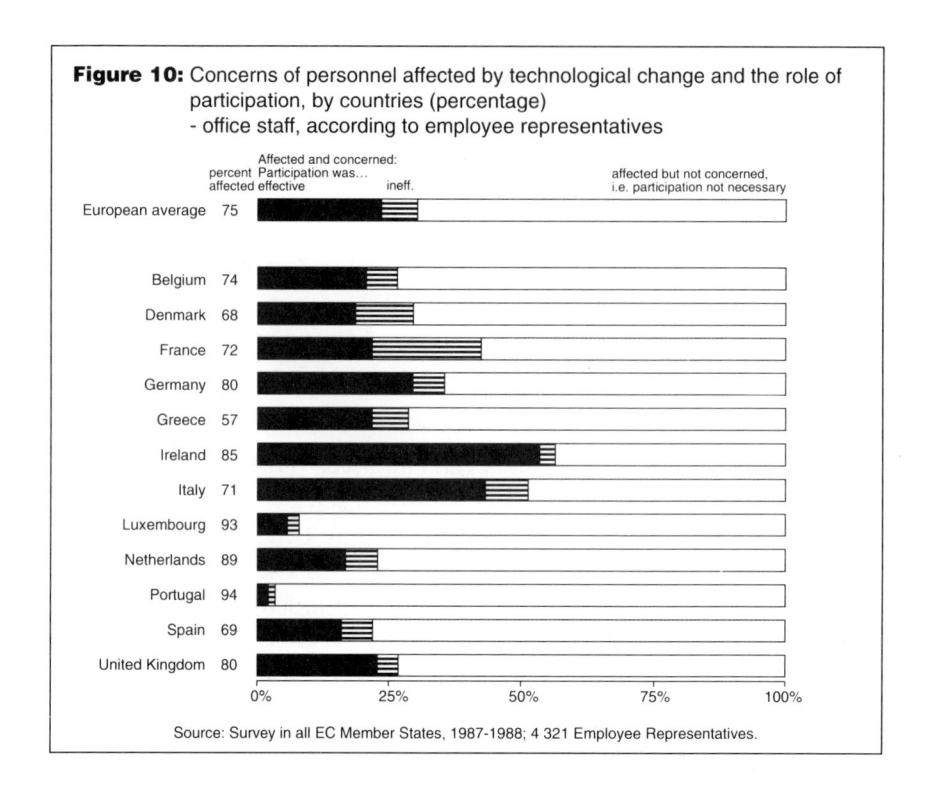

Figure 10: Concerns of personnel affected by technological change and the role of participation, by countries (percentage) - office staff, according to employee representatives

Country	percent affected	Affected and concerned: Participation was effective / ineff.	affected but not concerned, i.e. participation not necessary
European average	75		
Belgium	74		
Denmark	68		
France	72		
Germany	80		
Greece	57		
Ireland	85		
Italy	71		
Luxembourg	93		
Netherlands	89		
Portugal	94		
Spain	69		
United Kingdom	80		

Source: Survey in all EC Member States, 1987-1988; 4 321 Employee Representatives.

As to the value of participation in helping to alleviate these worries, managers generally have less faith than their employee representative counterparts — again particularly so in **Italy, France** and **Germany.** A majority of employee representative respondents in all Member States view participation positively, whereas managers in **France** are much more sceptical.

5.8.3 Secretarial Staff

Fewer secretarial staff were affected by technological change than was the case with middle managers and office staff. This might be because of the status attached by managers to the value of having a personal secretary, and hence the marked resistance on the part of managers to fully utilise the full potential of new information technology insofar as it affects their personal secretaries.[8] Again, management respondents in France reported that secretarial staff were concerned about new information technology to a greater extent than in other countries. In contrast with the responses of employee representatives, managers in most EC countries believed that fewer secretarial staff were concerned about new information technology.

Figure 11: Concerns of personnel affected by technological change and the role of participation, by countries (percentage)
- secretarial staff, according to managers

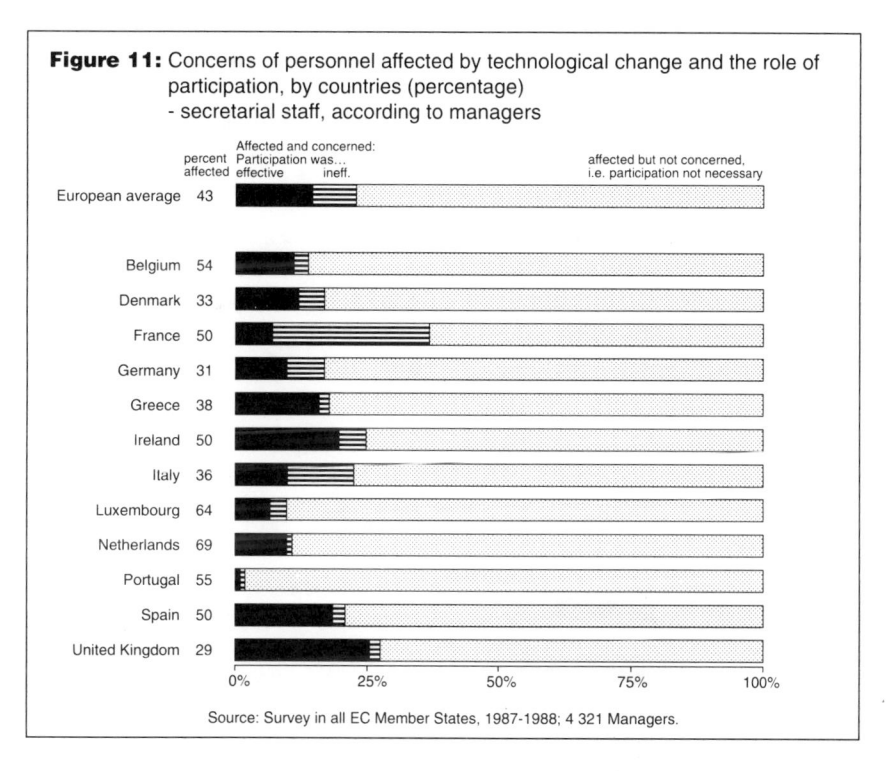

	percent affected	Affected and concerned: Participation was... effective / ineff.	affected but not concerned, i.e. participation not necessary
European average	43		
Belgium	54		
Denmark	33		
France	50		
Germany	31		
Greece	38		
Ireland	50		
Italy	36		
Luxembourg	64		
Netherlands	69		
Portugal	55		
Spain	50		
United Kingdom	29		

Source: Survey in all EC Member States, 1987-1988; 4 321 Managers.

Figure 12: Concerns of personnel affected by technological change and the role of participation, by countries (percentage)
- secretarial staff, according to employee representatives

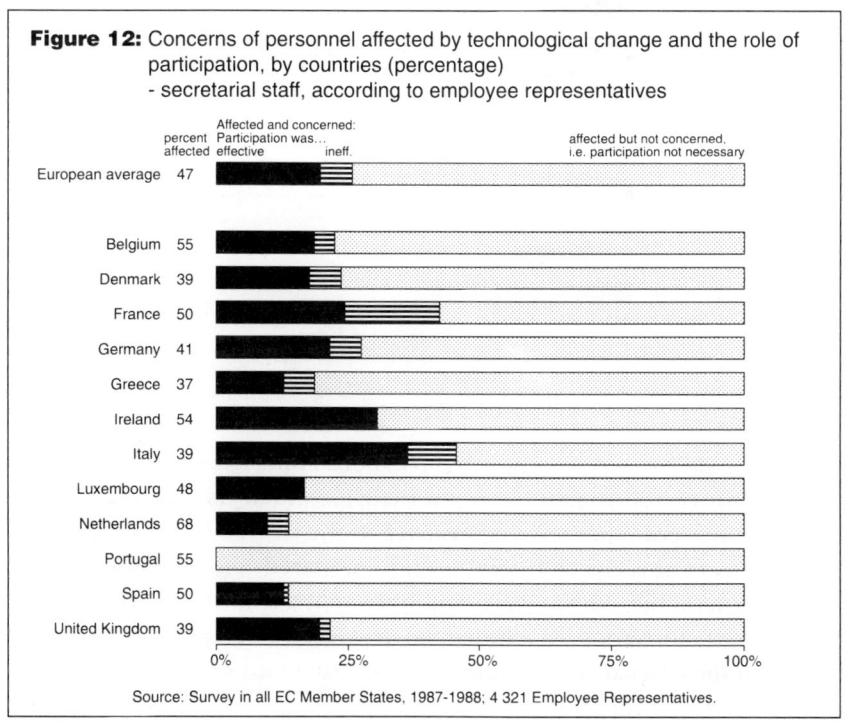

	percent affected	Affected and concerned: Participation was... effective / ineff.	affected but not concerned, i.e. participation not necessary
European average	47		
Belgium	55		
Denmark	39		
France	50		
Germany	41		
Greece	37		
Ireland	54		
Italy	39		
Luxembourg	48		
Netherlands	68		
Portugal	55		
Spain	50		
United Kingdom	39		

Source: Survey in all EC Member States, 1987-1988; 4 321 Employee Representatives.

How is participation valued by both sides in different countries? Participation is once again seen as effective in dealing with employees' problems by employee representatives rather than by managers — particularly in **France** and **Italy**. Very positive views on the value of participatory machinery is taken by both sets of respondents in two particular countries — **Spain** and the **United Kingdom**.

5.8.4 Shopfloor skilled workers

Just under half of all shopfloor skilled workers at the aggregate European level are reported by both sides to be affected by technological change. It is particularly notable that in the **Netherlands** around nine out of every ten such workers were reported to be affected.

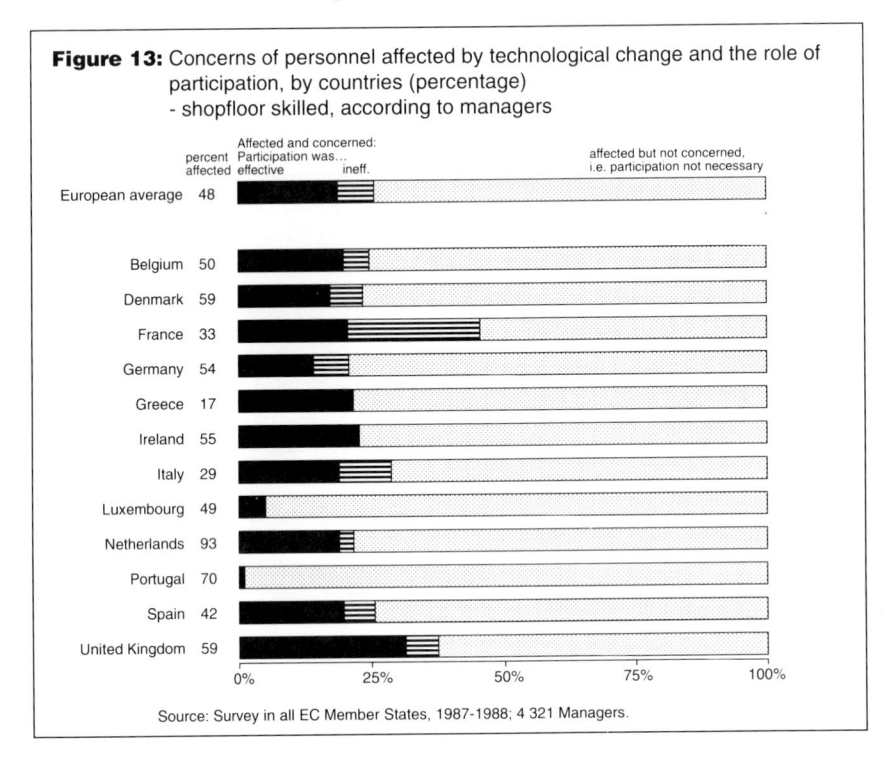

Figure 13: Concerns of personnel affected by technological change and the role of participation, by countries (percentage)
- shopfloor skilled, according to managers

Country	percent affected
European average	48
Belgium	50
Denmark	59
France	33
Germany	54
Greece	17
Ireland	55
Italy	29
Luxembourg	49
Netherlands	93
Portugal	70
Spain	42
United Kingdom	59

Source: Survey in all EC Member States, 1987-1988; 4 321 Managers.

Shopfloor skilled workers are slightly more concerned about the effect of new information technology on their jobs than the secretarial staff. A similar pattern emerges with respect to the degree of concern felt by employees in this occupational group compared with the white-collar occupations so far considered; employee representatives generally state that there is more concern felt than do managers. Both sets of respondents clearly indicate that two countries — **France** and **Italy** — again have the highest percentage of this occupational group who are worried about

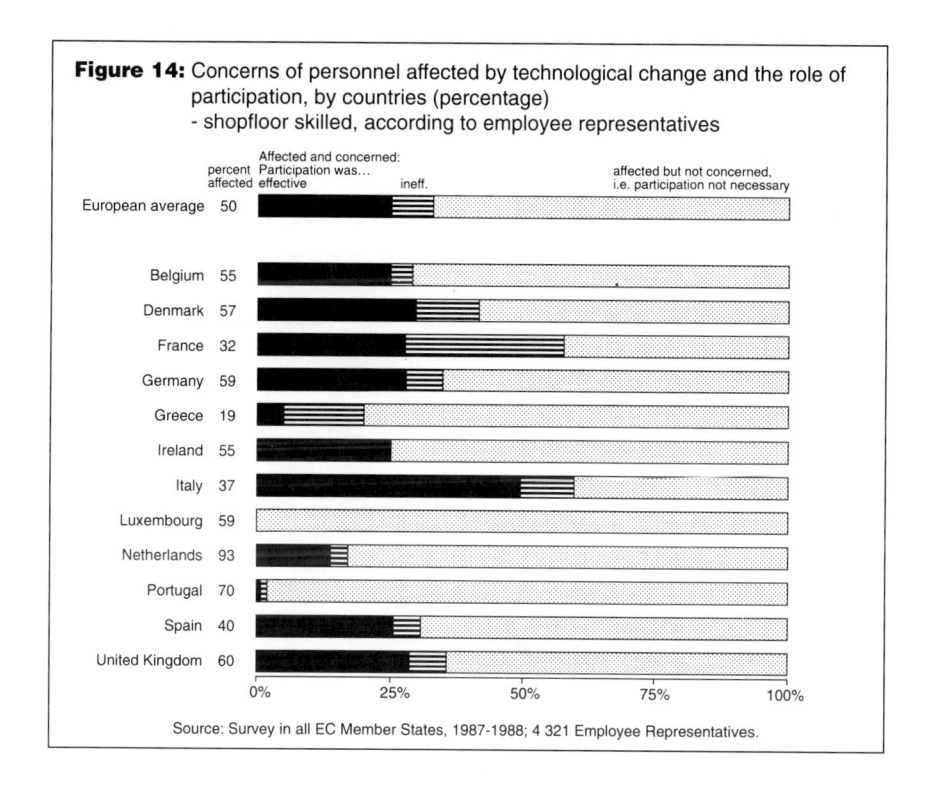

Figure 14: Concerns of personnel affected by technological change and the role of participation, by countries (percentage)
 - shopfloor skilled, according to employee representatives

	percent affected	Affected and concerned: Participation was... effective / ineff.	affected but not concerned, i.e. participation not necessary
European average	50		
Belgium	55		
Denmark	57		
France	32		
Germany	59		
Greece	19		
Ireland	55		
Italy	37		
Luxembourg	59		
Netherlands	93		
Portugal	70		
Spain	40		
United Kingdom	60		

Source: Survey in all EC Member States, 1987-1988; 4 321 Employee Representatives.

new information technology, closely followed by the **United Kingdom**; the percentage for **Portugal** again almost approaches zero.

How is participation seen by both sides as a means of dealing with the issues posed by technological change for skilled workers on the shopfloor? Employee representatives once again place more value on the role of participation than managers, although the difference between the two sides is not so pronounced with this occupational group. Indeed, it might be that managers generally see participation as more appropriate for manual workers than for white-collar workers and managers. It is notable that employee representatives in **Greece** are more sceptical about the value of participation than are managers. **France** showed the lowest support for participation by both sides.

5.8.5 Shopfloor semi-skilled and unskilled workers

A very similar pattern emerges for semi-skilled and unskilled shopfloor workers as for their skilled counterparts, except that they are marginally more concerned about the effects of new information technology on their jobs. According to both sets of respondents, over half of these workers are affected by technological change, with the **Netherlands** again having

Figure 15: Concerns of personnel affected by technological change and the role of participation, by countries (percentage)
- semi-, non-skilled, according to managers

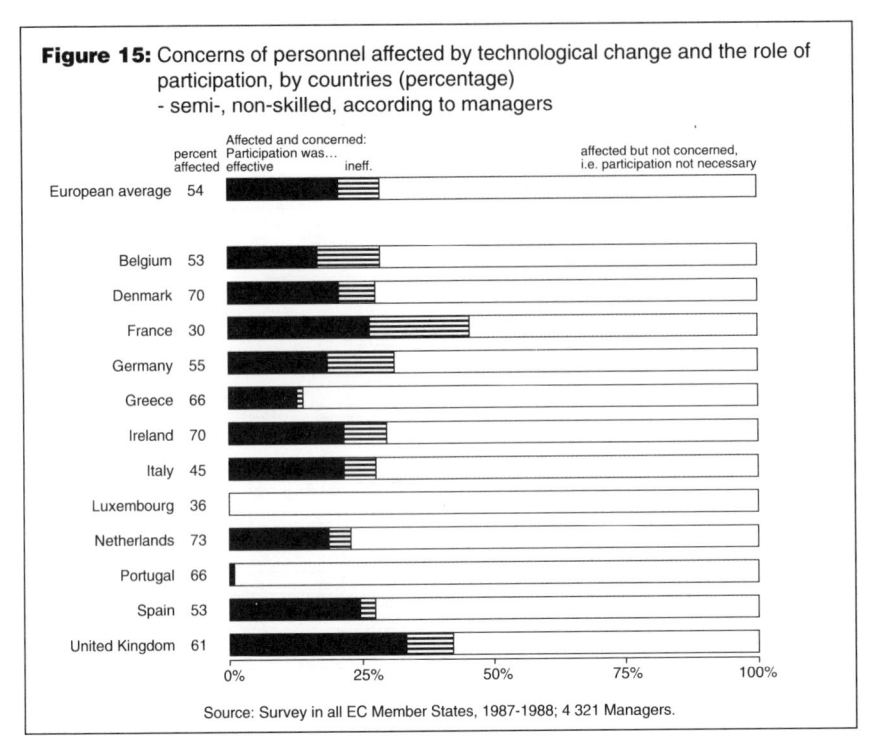

Country	percent affected
European average	54
Belgium	53
Denmark	70
France	30
Germany	55
Greece	66
Ireland	70
Italy	45
Luxembourg	36
Netherlands	73
Portugal	66
Spain	53
United Kingdom	61

Source: Survey in all EC Member States, 1987-1988; 4 321 Managers.

Figure 16: Concerns of personnel affected by technological change and the role of participation, by countries (percentage)
- semi-, non-skilled, according to employee representatives

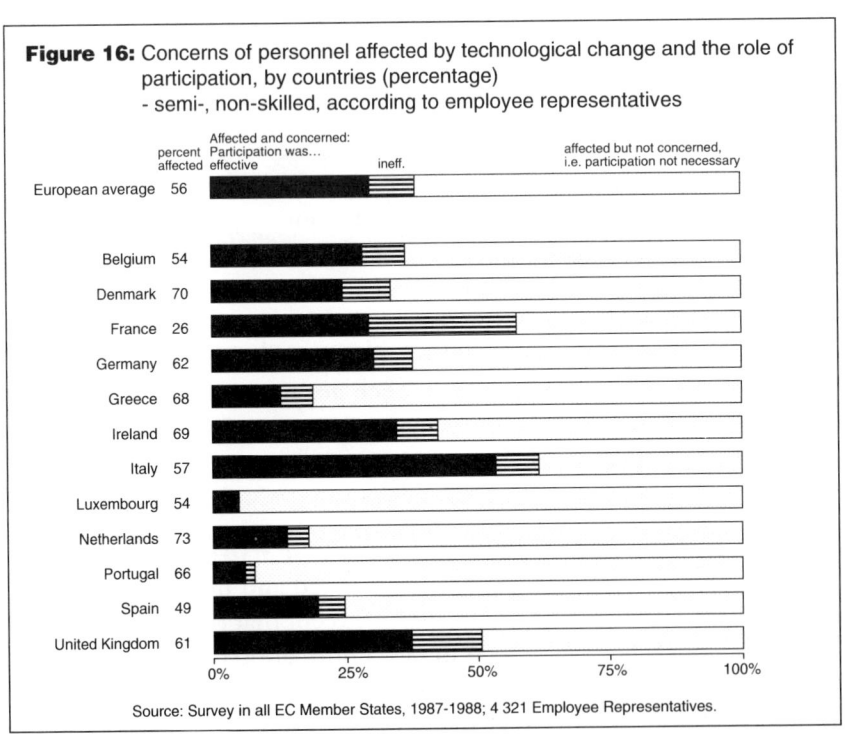

Country	percent affected
European average	56
Belgium	54
Denmark	70
France	26
Germany	62
Greece	68
Ireland	69
Italy	57
Luxembourg	54
Netherlands	73
Portugal	66
Spain	49
United Kingdom	61

Source: Survey in all EC Member States, 1987-1988; 4 321 Employee Representatives.

workers in this category who are most affected. According to managers, **France,** and the **United Kingdom** have the highest percentages of employees concerned about technological change; **Portugal** again registers a minuscule percentage. Employee representatives agree with their management counterparts that semi-skilled and unskilled workers in **France** and the **United Kingdom** have been affected most, but employee representatives in **Italy** also claim that such workers have been very much affected.

As for support of both sides for participation, once again management support is marginally higher for this group of manual workers than for white-collar workers, but less so generally than is the case for employee representatives.

5.9 General

There is one point that can be made about the country differences with respect to the effects of technological change on the jobs of various occupational groups. Whilst it seems to be the case that most employees support new information technology and are not overly concerned about its effects on their jobs, two particular countries stand out across all occupational categories — **France** and **Italy.** In these two countries significantly more employees appear to be worried about technological change. This might well be due to the low trust relations that exist between management and trade unions in those two countries. In both countries, there is a history of suspicion between the two sides of industry, and the major trade union confederations have an ideological antipathy to management. Management, in turn, tend to be more confrontational. The concern felt by employees about technological change in **France** and **Italy** may well be not so much a concern about new information technology as such, but a reflection of their suspicions of management policies.

The second point to note is that, generally speaking, participation is seen as an effective means of resolving problems arising from technological change in all countries and across occupations.

Finally, there is some evidence from the survey to suggest that managers' support for participation as a means of dealing with problems associated with technological change seems to be marginally higher for manual workers rather than for white-collar employees. This might be a reflection of the higher degree of unionisation for the former category.

5.10 Conclusions

Our survey findings provide support to those who argue that new information technology has had a beneficial effect on the jobs of those employees in the forefront of technological change. When it has been introduced it has encountered little resistance and few employees express fears about its adverse effects. There is little evidence in our survey results to suggest that jobs have disappeared as a result of the introduction of new information technology into European workplaces; on the contrary, what evidence we do have points to the opposite conclusion, namely that if anything, more jobs have been created as a result. In all these respects our survey findings are in line with those of Daniel's major workplace industrial relations survey on new information technology in the **United Kingdom** and other surveys, including those from the OECD and in **Germany**.

We have no methodologically sound means of plotting the complex nature of employment flows that clearly took place in the enterprises in our survey; it is clear that many jobs must have been altered in form, large numbers of workers were transferred to other jobs within their companies, and many jobs must have been subsumed into other occupations. Nor do we have any means of knowing if technological changes in the enterprises caused major job losses elsewhere or if it enabled managers to avoid taking on new employees in circumstances where they would otherwise have done so.

It also seems that participation provides an effective means of resolving the problems that employees in various occupational groups may have about the impact of new information technology on their jobs. This result isby no means surprising but it lends support to those who believe that such participatory machinery is essential in the management of change in companies.

We would certainly not claim that our results are not open to challenge. As many readers will be only too aware, our survey was dealing with enterprises which had been screened to ensure that only those with some form of formal employee representative machinery took part in the survey. In this respect it could legitimately be argued that our sample was biased in favour of firms which were prosperous, had good industrial relations and where employees saw investment in new information technology by management as a belief in future company prosperity. It might be also be argued that the respondents in our survey were more likely to take a favourable view about the impact of new information technology simply because they were recounting experiences in the past and any apprehensions they may initially have had did not materialise. Nevertheless, we can only repeat one of the central findings of this part of the survey — namely **that new information technology appears to have created jobs rather than destroying them, at least in the sectors in which the survey was carried out.**

Finally, new information technology has had a pervasive, if generally beneficial effect, on jobs throughout European enterprises during the 1980s where large numbers of employees from all occupational groups felt its effects. In this respect, the early predictions that were made about new information technology being a truly "heartland" technology have certainly been realised. It seems from our results that new information technology tends to have a large impact at the top of organisations, with large numbers of managers and office staff affected. This is not to say that other jobs in the occupational hierarchy have not been affected - quite the contrary; however, it seems that by the time new information technology reaches these other occupational categories, other organisational factors tend to mitigate its impact.

NOTES

1. W.W. Daniel, *Workplace Industrial Relations and Technical Change*, Frances Pinter, London, 1987.
2. Organisation for Economic Co-operation and Development, *New Technology in the 1990s: a socio-economic strategy*, Report of a Group of Experts on the Social Aspects of New Technologies, Paris, 1988.
3. Bundesminister für Forschung und Technologie (ed.), *Arbeitsmarktwirkungen moderner Technologien,* Bonn, 1988, p. 34 and pp. 206-209.
4. W.W. Daniel, *Workplace Industrial Relations and Technical Change,* DE/ESRC/PSI/ACAS, Frances Pinter, 1987.
5. Colin Gill, *Work, Unemployment and the New Technology,* Polity Press, Oxford, 1986.
6. There is an extensive literature on this trend towards more simplification of management structure as new technology is introduced into organisations. See in particular, J.Child, New Technology and developments in management organisation, *Omega,* vol.12, no 3, 1984, p.218.; C.Handy, *The Future of Work,* Basil Blackwell, Oxford and New York, 1984.
7. One other problem of the Irish sample of respondents is that the sample was compiled from the Dunn and Bradsheet list of the top 500 Irish companies; in Ireland this directory is regarded by some researchers as a poor source of information for survey purposes.
8. Hazel Downing, Word Processors and the oppression of women, in T.Forester (ed.), *The Micro-electronics Revolution,* Basil Blackwell, p. 275, 1980.

Chapter Six

Participation in Training

6.1 Introduction

The creation of the Single Market in the European Community will lead to a great deal of restructuring of industry. All Member States will need to respond to radical changes in the labour market and employees in European enterprises will have to adapt to a higher level of skills. There is therefore a need for all the Member States to adopt policies on education and training which, while respecting the different systems that exist throughout the Community, needs to be able to respond to the pressures of the 1990s — particularly increased global competition. The labour market will be increasingly European in character and in order to take advantage of the opportunities which will be created as a result of the setting up of the Single Market, changes in attitudes will be required throughout industry and the social partners will need to have a much better knowledge of other Member States.

During the 1980s the Member States increasingly recognised the importance of the challenges that lay ahead. There were a large number of reforms and reviews which were designed to extend, diversify and improve the level of training provision. The European Commission has a duty to develop a common vocational training policy under Article 128 of the Treaty of Rome. So far, the Commission has updated the 1963 Decision which set out the basic principles of a common vocational training policy and has made commitments to prepare an instrument as part of the Social Charter (see Chapter 3) proposing a right of access to training, to make proposals for rationalising existing Community-level training programmes and to speed up action on establishing comparability of training qualifications.

It is clear that training has a fundamental role to play in strengthening the competitive position of European firms and in reviving growth. As the OECD Intergovernmental Conference on Further Education and Training of the Labour Force concluded:[1] "Efficient diffusion and application of advanced technologies depend as much on the people who use them and on a wide spectrum of collateral social, organisational, economic, legal and even cultural changes. The skills and competences of the workforce, from production workers to senior management, are proving to be crucial determinants of productivity and competitiveness". This is particularly true as the Single Market develops. The Single Market will reveal new requirements as regards training or will accentuate others. In short, because training, particularly continuing training, is a way of creating a European pool of skills capable of intensifying the positive effects of the single market, it is one of the key priorities of the 1990s for European firms. Moreover, there is a great deal of consensus between the two sides of industry about the importance of training. Given this consensus between the two sides about the importance of training, the success of training in the future depends on a partnership not only between governments, but also between management and employees.

As we saw in Chapter 3 of this Report, there have been a number of Joint Opinions between UNICE and CEEP on the one hand, and the ETUC on the other which affect training. The first two of these were those on "Training and Motivation" and "Information and Consultation" which wereconcluded at Val Duchesse in 1987. Since then there have been four other Joint Opinions affecting training:

a) "The Creation of a European Occupational and Geographical Mobility Area and Improving the Operation of the Labour Market in Europe" which was concluded in February 1990;

b) "New Technologies, Work Organisation and Adaptability of the Labour Market", (January 1990);

c) "Education and Training," (June 1990);

d) "The Transition from School to Adult Working Life", (November 1990).

There are two important themes which run through all these four new Joint Opinions. The first is that training is assigned a fundamental role throughout the "...working life of employees in order to promote a workforce which is better trained, more motivated, more mobile and able to adapt to change and the new qualification requirements..."[2] The second theme which runs throughout these four Joint Opinions is the need for information and consultation on training programmes.

6.2 Past Participation in Training at the European Level

Given the importance of the training issue for the success of European enterprises in the 1990s, what does our survey reveal about the degree of participation in training generally?

There is a great deal of consensus between the two sets of respondents about the degree of participation that existed in European enterprises at the time the survey took place and we have combined the responses of both sides together at the aggregate European level as in figure 17 overleaf.

The results show that only one in two employee representatives were either consulted or had a stronger form of involvement in training. Moreover, about one in five employee representatives had no involvement whatsoever in training; 29% of employee representatives were merely

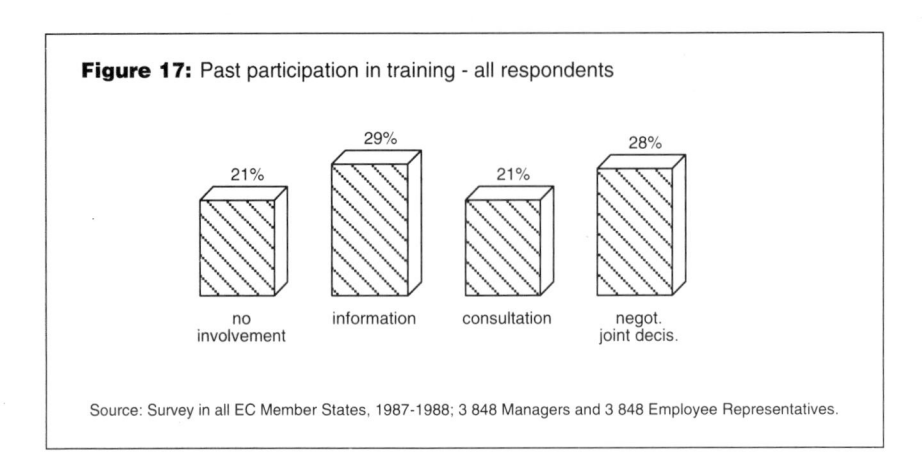

Figure 17: Past participation in training - all respondents

- no involvement: 21%
- information: 29%
- consultation: 21%
- negot. joint decis.: 28%

Source: Survey in all EC Member States, 1987-1988; 3 848 Managers and 3 848 Employee Representatives.

provided with information about the training that took place in their companies. These results do not augur well for the future, especially in view of the fact that training occupies a high place on the agenda of European companies in the 1990s and that its importance is felt by both sides of industry.

6.3 Inter-country comparisons of participation in training

In this section of the report we explore the forms of participation in training that were in evidence at the time the survey interviews were carried out in 1987-88. Secondly, we consider the kind of participation that managers and employee representatives would prefer in the future. Throughout this section we seek to account for the differences in participation in training that exist between all the 12 EC Member States by using our explanatory framework which was outlined earlier in Chapter 4 and comment on the results in the light of the various Commission Directives and proposals which are likely to be implemented as part of the Social Action Programme (see Chapter 3).

Draft Figures 18 and 19 give a picture of the degree of participation in training that existed as reported by managers and employee representatives at the time the interviews were carried out.

The most striking observation that can be made is that managers and employee representatives are in fairly close agreement about the levels of participation in existence, although managers tend to give a marginally more optimistic picture than their employee representative counterparts. **Denmark** and **Germany** have the greatest degree of negotiation and/or

PARTICIPATION IN TRAINING

Figure 18: Past participation in training in the EC states - Managers

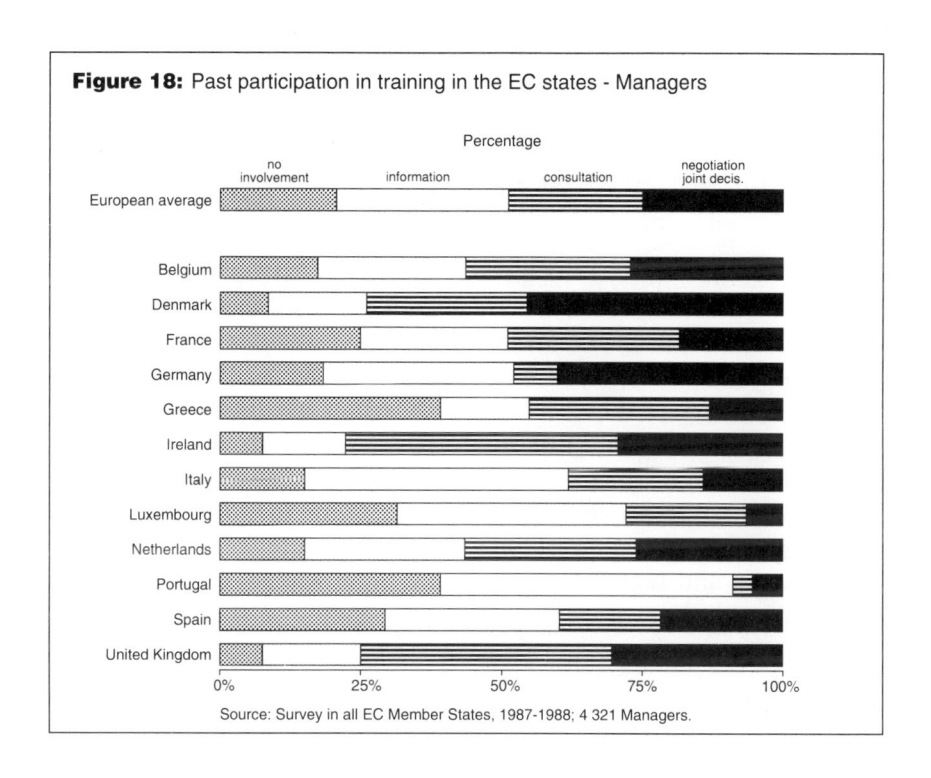

Percentage — no involvement / information / consultation / negotiation joint decis.

European average, Belgium, Denmark, France, Germany, Greece, Ireland, Italy, Luxembourg, Netherlands, Portugal, Spain, United Kingdom

Source: Survey in all EC Member States, 1987-1988; 4 321 Managers.

Figure 19: Past participation in training in the EC states - Employee Representatives

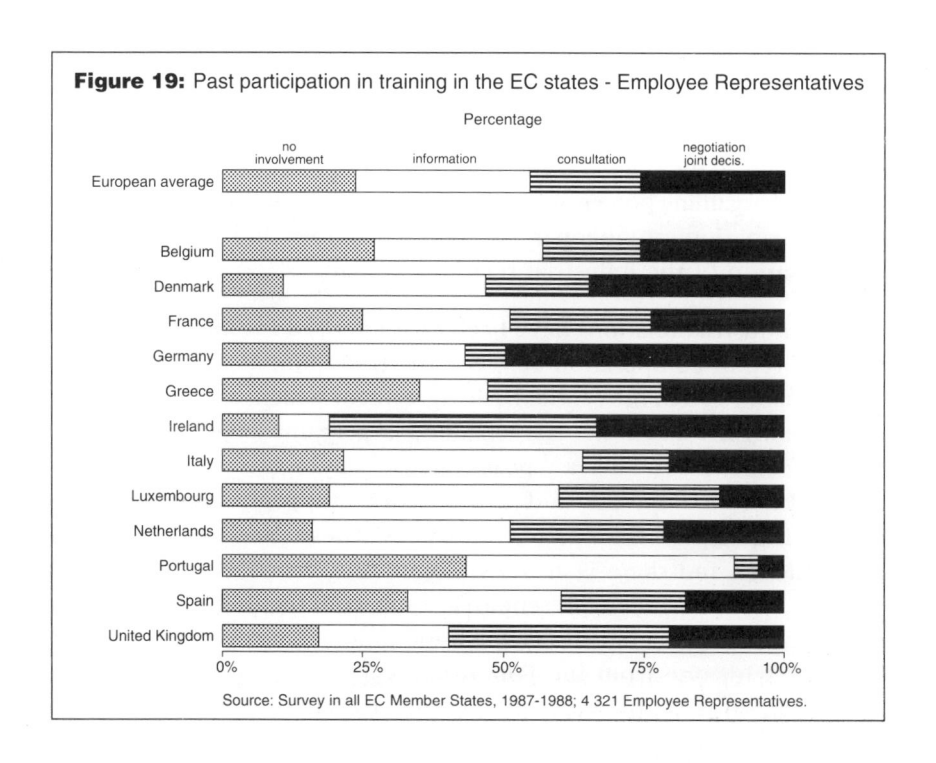

Percentage — no involvement / information / consultation / negotiation joint decis.

European average, Belgium, Denmark, France, Germany, Greece, Ireland, Italy, Luxembourg, Netherlands, Portugal, Spain, United Kingdom

Source: Survey in all EC Member States, 1987-1988; 4 321 Employee Representatives.

joint decision making in training although **Denmark** has a greater degree of consultation than is the case in **Germany**. These two countries are followed by a middle-ranking group of countries — **Ireland, Belgium, The Netherlands,** and the **United Kingdom.** The lowest group of countries includes **Spain, France, Greece, Italy, Luxembourg** and **Portugal.**

Despite the high degree of consensus that exists between the two sides of industry on the importance of training, it is depressing to note the high degree of "no involvement" by employee representatives in training. Just under one in four employee representatives at the aggregate European level (Figure 19) report they have no part to play whatsoever in training. In certain countries — **Portugal, Greece** and **Spain** — there a particularly high level of no involvement in training with around four out of every ten employee representatives stating that they had no part to play in training. In contrast, only one in ten employee representatives in **Denmark** and **Ireland** reported that there was no involvement on their part.

6.4 Explaining inter-country differences in training

How can these differences be explained? According to managers in **Denmark**, this country has the highest degree of negotiation and/or joint decision-making (see Figure 18); Danish employee representatives, however, are much more sceptical (Figure 19). If we consider the case of **Denmark** first and apply our explanatory factors, it is evident that all five factors are favourable in promoting participation -managements' dependence on the skills and co-operation of the workforce, management style, the bargaining power of organised labour, regulations which provide for participation on the part of the workforce and the degree of centralisation of the industrial relations system.

Danish industrial relations have been characterised for many years by a high degree of co-operation between the trade unions and employers. This is manifested in the central Basic Agreement which, apart from minor changes in 1981 and 1986, has remained largely unchanged since 1973; this agreement contains a "peace clause" by which the two sides undertake to co-operate and avoid work stoppages. In companies employing more than 35 employees there are 'co-operation committees'[3] and there is also a Co-operation Board at national level which consists of three representatives each from the Danish Employers' Confederation (DA) and the major trade union confederation, the LO, together with one from the Foremans' Association. The national Co-operation Board provides guidance to firms concerning the implementation of the Basic Agreement and the role and function of co-operation committees at plant level. In addition, since much of Danish industry relies on the manufacture of specialist quality products, there is an emphasis placed on using new information technology to increase competitiveness, to enhance product quality and to improve customer service. In such circumstances, the training of employees to work with new information technology is given a high priority.

Relations between employers and trade unions in Denmark have long traditions of co-operation and this has been underpinned by a system of tri-partite consultation and state support for jointly agreed goals between the two parties. Trade unions are accepted by management as partners in the planning of technological change and there is a stock of necessary goodwill between the two sides of industry. There is thus a high dependence by employers on the skills and co-operation of their workforce.

Trade unions in Denmark are well organised and cover almost the entire labour market. Trade unions are accepted as useful partners in making the fullest use of new information technology. The Danish trade unions in turn are committed to technological change and there is a high trade union density in Danish companies. The Co-operation Committees work well, and there is a substantial degree of co-determination on many technological matters where both sides strive for agreement. In comparison to **Germany**, there is much less legislation, but nevertheless the more voluntaristic features of the Danish industrial relations system serve to ensure that issues concerned with new information technology are regularly discussed between both sides of industry. The Danish government has traditionally placed much emphasis on training. Through legislation the State has provided the framework for the organisation and direction of vocational training. The various areas of training are the responsibility of different ministries, as well as State and community administrations. Training systems are guided and shaped at all levels by a well-developed system of councils and committees. A cardinal feature of training in **Denmark** is that participation of both employers and employee organisations is a basic principle for all bodies concerned with training.

Finally, **Denmark** has perhaps the most centralised industrial relations system in the Community where both employers and unions are able to assert authority over their respective affiliates. All five of our explanatory factors are therefore favourable, and they explain why **Denmark** has the highest degree of participation by employee representatives in training matters.

According to managers, **Germany** has the second highest degree of participation in training (Figure 18), whereas if we consider the views of employee representatives (Figure 19), the positions between **Denmark** and **Germany** are reversed! Most of our explanatory factors are generally favourable in **Germany**, although not to the same extent. In contrast to **Denmark**, **Germany** has a much more legalistic system and there is much more diversity from one company to another in management attitudes towards participation. In the larger companies management is more co-operative in its style, whilst in the smaller companies in tends to be paternalistic in its nature.

There is a long tradition of legislation in respect of vocational training in **Germany** which stretches back to the pre-war period.[4] The numerous regulations for workplace training were consolidated into the Law for Vocational Training (BBiG) in 1969. Other legislation affecting training includes the Law for the Promotion of Employment (AFG, 1969); The Regulation for the Competence of Trainers (AEVO, 1972); the Law for the Constitution of Companies (1972); and the Law Regulating Training Places (APIFG, 1976). The Law on the Constitution of Companies gives works councils the right to participate in all decisions concerning training.

Germany also has a much less centralised industrial relations system, with only a degree of centralisation at sector level. Moreover, managers in German companies are known to be opposed to any further improvements in works council legislation which would have the effect of giving more influence to employee representatives in technological change. Although German trade unions are well resourced, the absence of a formal presence at workplace level where major decisions on new information technology are taken is a disadvantage to the unions, although unions there have used the works council system as a *de facto* union structure.

The nature of participation in German companies has undergone substantial change during the 1980s; these changes are the result of a number of management-sponsored initiatives on participation. The new management initiatives deal with participation of an instrumental character, whose aim is to mobilise the productive and motivational resources of the employees, or, to be more exact, the resources of a privileged core labour force.

As Otto Jacobi and Walther Müller-Jentsch[5] point out, the ability of management to practice the involvement of individual workers exclusive of formal institutionalised representation depends on the influence of trade unions on works councils within the firm. There is evidence that

German managers in expanding industries with a high percentage of white-collar employees are pursuing flexible forms of participation of a consultative nature which is centred on individual employees. In situations where management cannot ignore the institutionalised workers' representation, it will attempt to use it as an instrument for its own purposes. Since our survey only covers companies which have some form of formal participation mechanisms and is only concerned with institutionalised representation, these major changes in German companies — important as they are — will not be apparent in our data.

Surprisingly, the **United Kingdom** has a comparatively high level of participation in training with a particularly high degree of consultation; the level of "no involvement" in that country is the lowest of all the European Community countries. Three out of every five British managers claim that participation in training either consists of consultation or negotiation/joint decision-making; employee representatives, however, tend to take a more sceptical view about participation levels. As regards participation in training, our explanatory factors are more favourable than they are in other areas of technological change.

Management in the **United Kingdom** has a fairly high dependence on the skills and co-operation of the labour force, particularly in the implementation stage of technological change where training is of great importance. The trade unions in Britain are seeking a new role to extend their influence to counteract the employer offensive that took place during the 1980s. Several of the major unions are seeking to equip their members with new skills — whether through participating in the reform of training schemes or, in some cases, through the provision of training by the unions themselves. Given the severe shortage of skilled workers in Britain, it is not surprising that training is an area where employers are prepared to seek a joint approach with employee representatives in order to overcome their skill shortage problems.

Despite the fact that the **United Kingdom** has a very decentralised industrial relations system — a factor which appears at first sight to be unfavourable to participation — in the training area this is tempered by the dependence that management at company level has in relying on the co-operation of employee representatives in ensuring that the necessary skills are available when new information technology is introduced into the workplace. All this should be qualified by noting that although there is a comparatively high involvement in training by employee representatives in the **United Kingdom**, since the survey was carried out the British government removed union representation from the training bodies and instituted essentially employer-run Training Enterprise Councils.

Earlier in this chapter it was noted that the results for **Ireland** must be interpreted with some caution because of the small sample of respondents. Some of the explanatory factors for **Ireland** are unfavourable, although there is evidence that management there is receptive to the outcome of "National Understandings" between themselves, the government and the trade unions. The National Apprenticeship Board (An Cheard Chomhairle, AnCO), apart from the supervision of apprenticeship skills, has the task of promoting or providing training for all industrial skills, and its management is tri-partite in character with representatives from government, employers' organisations and trade unions.

The industrial relations system in **Ireland** is also becoming more centralised and Irish trade unions have more bargaining power than their British colleagues. While there is legislation promoting participation in Irish state-owned industries, these industries were not covered by our survey.

Among the other middle-ranking countries, both **Belgium** and the **Netherlands** have legislation which gives consultation rights on training to works councils. However, the works council system in **Belgium** has a limited role in enabling employee representatives to influence issues concerned with technological change because such councils are essentially only advisory in character. The Belgian trade unions tend to rely either on clauses within collective agreements or on other legal provisions to exert their influence; trade union density in **Belgium** is high. In the **Netherlands** there are generally favourable factors in operation: there is a fairly centralised industrial relations system (although in the 1980s there has been increased scope for more decentralised bargaining, especially over working hours and training); a well-established works council system through which the law promotes worker participation (although the unions do not control the councils); and a high dependence by management on the skills and co-operation of employees. However, during the 1980s there has been a drastic decline in trade union density and bargaining power.

All the lower-ranking countries in the table have generally unfavourable factors in operation. **Spain, Portugal** and **Greece** all have legacies of a dictatorial past with all this means for inhibiting trade union development. In many of these countries management is generally hostile towards participation (**Greece, Portugal, Spain, France** and **Italy**) or the trade unions prefer to subsume technological change under collective bargaining rather than sacrificing their traditional oppositional role by involving themselves in works council arrangements. Even where there have been recent efforts within the wider trade union movement to forge closer links and promote a greater degree of unity (e.g. **Spain, Italy** and to a much lesser extent **France**), such efforts have had very little impact as yet on participation. In any case, while many of these lower-ranking countries have formal, institutional works council provisions which have emanated from legislation, it is often the case that such works council arrangements are inadequate in providing much influence to employee representatives in dealing with new information technology issues such as training.

6.5 Participation in various training issues

Respondents in those companies who reported that there *was* some degree of participation in training issues were invited to respond to a multi-response question which asked them to indicate one or more issues in which employee representatives played an influential role.

Figure 20: Participation in different training issues according to managers and employee representatives - multiple responses (percentage)

Managers: selection of partic. 22, timing and duration 30, contents of training 37, type of training 44
Employee represent.: selection of partic. 23, timing and duration 31, contents of training 36, type of training 48

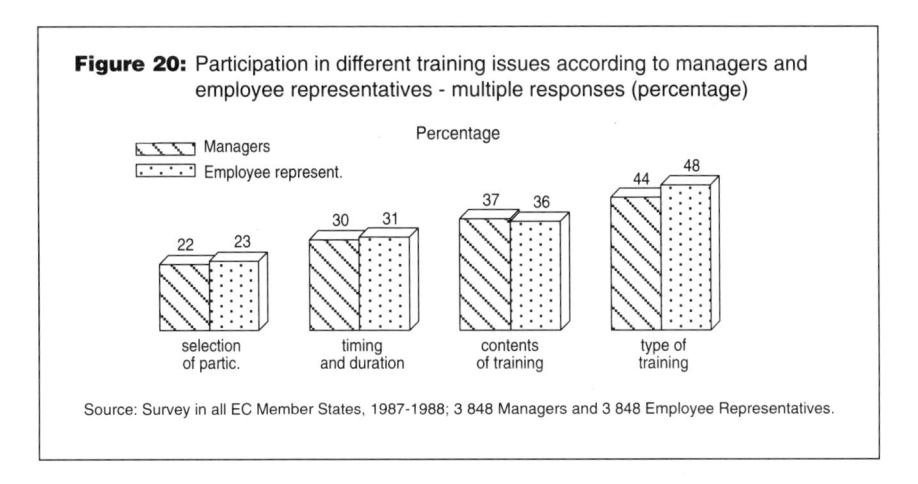

Source: Survey in all EC Member States, 1987-1988; 3 848 Managers and 3 848 Employee Representatives.

Figure 20 gives an indication of the importance of participation in four main training issues at the aggregate European level according to the responses of both sides in those enterprises where the respondents stated that there was some participation by employee representatives.

Figure 20 illustrates that there is much agreement between both sides in their responses to this question. It appears that employee representatives play a bigger role in decisions about the type of training to be done than any other issue. This is closely followed by participation in the contents of training programmes. Three out of every ten respondents state that participation occurs about the timing and duration of training programmes, while the selection of participants seems to be very much a matter of managerial prerogative.

6.6 Inter-country comparisons of participation in various training issues

6.6.1 Selection of participants

When new information technology is introduced into the workplace it frequently has the effect of eroding the traditional boundaries between occupations. Conflicts can often arise between different occupations when new information technology is introduced into the office or on the shopfloor. New information technology often requires a different range of skills on the part of the operator. The choice that is made as to who is selected to work with the new information technology is therefore often contentious, and much depends on the way management decides to use the new information technology.

What kind of employee representative involvement took place in this area in the different Member States?

Figure 21: Participation in different training issues according to managers and employee representatives

- Selection of participants — Percentage

Country	Managers	Employee repres.
European average	22	23
Belgium	17	20
Denmark	27	32
France	7	11
Germany	32	37
Greece	39	13
Ireland	17	12
Italy	18	20
Luxembourg	23	35
Netherlands	23	22
Portugal	11	9
Spain	14	11
United Kingdom	30	25

Source: Survey in all EC Member States, 1987-1988; 4 321 Managers and 4 321 Employee Representatives

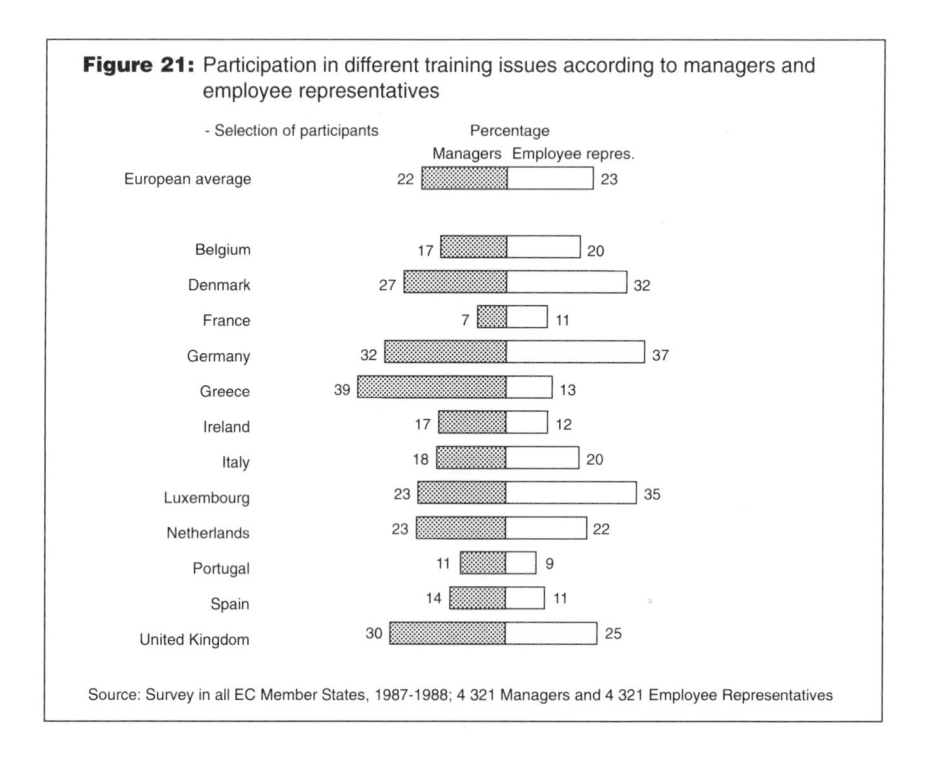

Figure 21 above gives a breakdown by country of participation in the selection of participants for training in new information technology. The figure shows that the European average conceals a great deal of variation between Member States not only in relation to the amount of participation but also differences between the responses of both sides in each Member

State. It is particularly notable that there is little participation in **France** and **Portugal** and managers in **Greece** claim that there is more participation than employee representatives there accept. In general, the selection of participants for training seems to be an issue of managerial prerogative.

6.6.2 Timing and duration

This area of training is characterised by marginally more participation than the selection of participants, but again the European averages conceal much diversity (Figure 22). **Portugal** has figures which are the highest, thus reversing its position under 6.6.1 above. This is probably a reflection of the extremely legalistic industrial relations system there. It is also worth noting that there is a great deal of difference between the managers' and employee representatives' responses in the case of **Spain**, with over nine out of ten employee representatives claiming that participation exists,

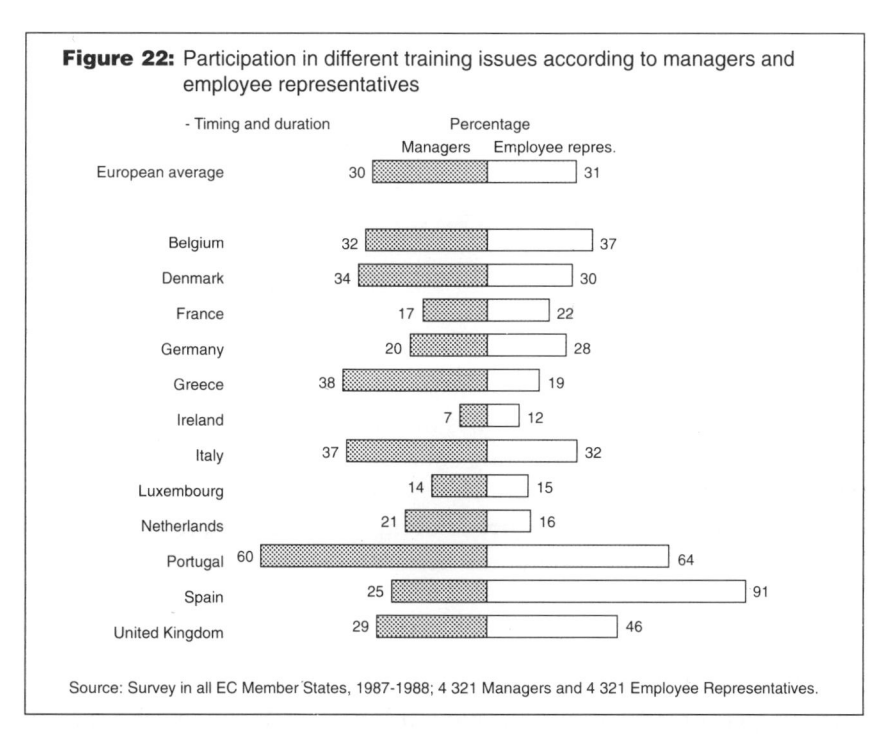

Figure 22: Participation in different training issues according to managers and employee representatives

- Timing and duration — Percentage

	Managers	Employee repres.
European average	30	31
Belgium	32	37
Denmark	34	30
France	17	22
Germany	20	28
Greece	38	19
Ireland	7	12
Italy	37	32
Luxembourg	14	15
Netherlands	21	16
Portugal	60	64
Spain	25	91
United Kingdom	29	46

Source: Survey in all EC Member States, 1987-1988; 4 321 Managers and 4 321 Employee Representatives.

compared with only a quarter of management respondents. There are also differences between the responses of the two sides in the **United Kingdom**, with 46% of employee representatives claiming that participation exists compared with 29% of managers. **Ireland** has the lowest figure for participation in this issue of training, followed by **Luxembourg**.

6.6.3 Contents of training

Figure 23 shows that there is not so much diversity between the Member States in respect of this training issue, nor is there much difference between the responses of both sides in each country.

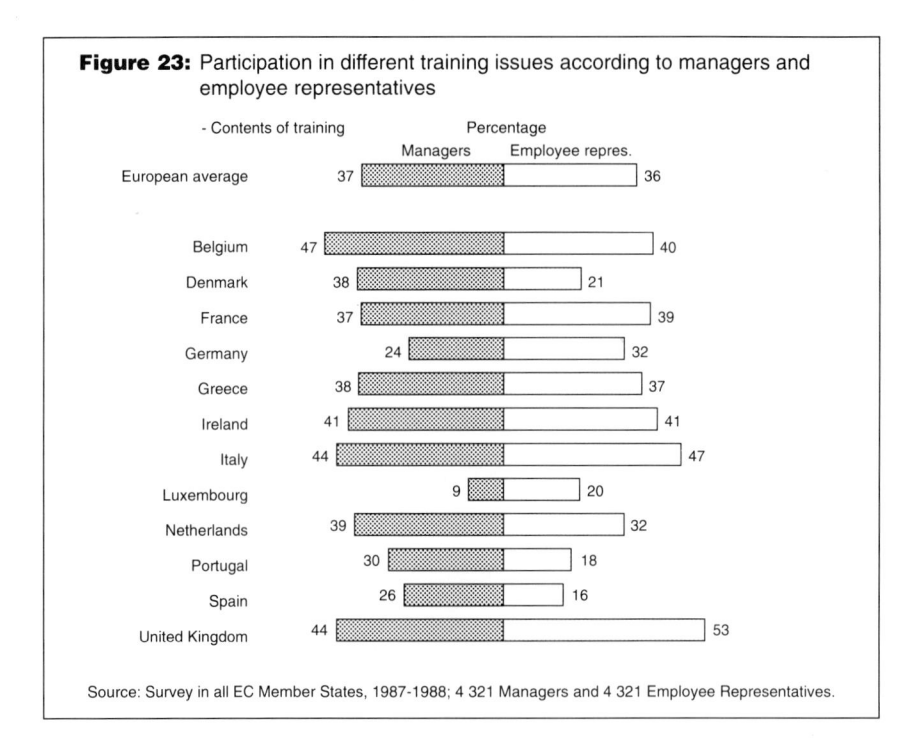

Figure 23: Participation in different training issues according to managers and employee representatives

- Contents of training — Percentage

Country	Managers	Employee repres.
European average	37	36
Belgium	47	40
Denmark	38	21
France	37	39
Germany	24	32
Greece	38	37
Ireland	41	41
Italy	44	47
Luxembourg	9	20
Netherlands	39	32
Portugal	30	18
Spain	26	16
United Kingdom	44	53

Source: Survey in all EC Member States, 1987-1988; 4 321 Managers and 4 321 Employee Representatives.

Participation levels are highest in **Belgium, Ireland, Italy,** and the **United Kingdom**, where over 40% of managers report that involvement occurs in decisions about training content; it is lowest in **Luxembourg, Spain** and **Portugal.** There are significant differences between the two sides of industry in **Luxembourg, Portugal,** the **United Kingdom,** and **Denmark**, where the differences between the two sides exceed 10%.

6.6.4 Type of training

In this aspect of training, which enjoys the greatest degree of participation of all training issues, the variations are again fairly minimal. It is somewhat surprising that **Germany** has the lowest figure for participation of all managers' responses; however, employee representatives there take a more positive view in their responses. The figures for **Greece** are particularly high.

There are fewer cases where the responses of managers differ greatly from

PARTICIPATION IN TRAINING

those of employee representatives; the differences exceed 10% only in the **Netherlands** and in **Germany**.

Figure 24: Participation in different training issues according to managers and employee representatives

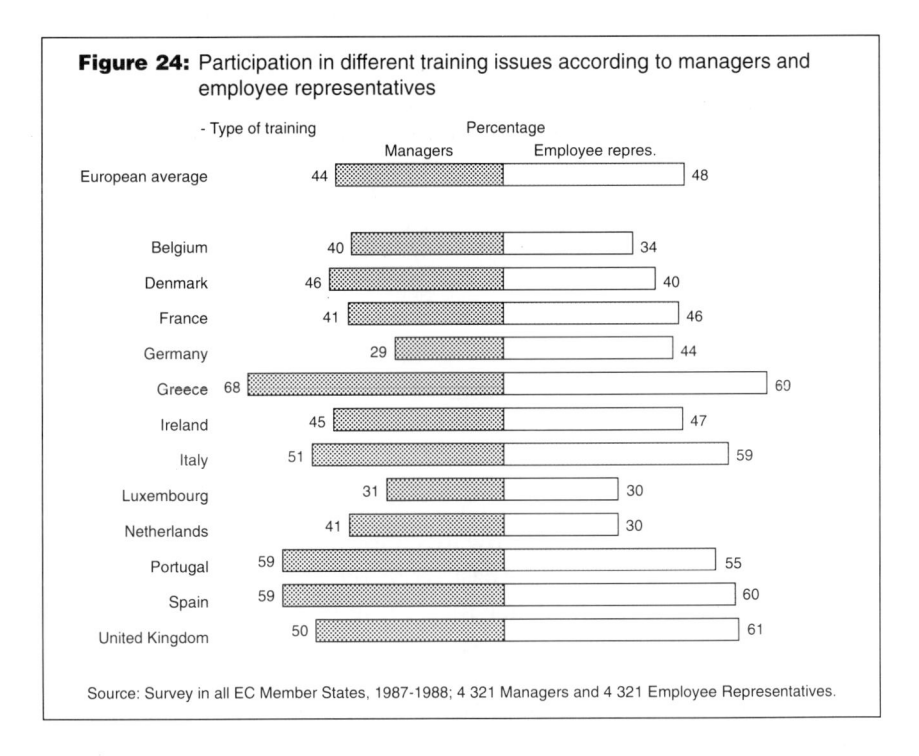

- Type of training	Managers	Employee repres.
European average	44	48
Belgium	40	34
Denmark	46	40
France	41	46
Germany	29	44
Greece	68	60
Ireland	45	47
Italy	51	59
Luxembourg	31	30
Netherlands	41	30
Portugal	59	55
Spain	59	60
United Kingdom	50	61

Source: Survey in all EC Member States, 1987-1988; 4 321 Managers and 4 321 Employee Representatives.

6.7 Assessing the potential for participation in training in the future

In our survey we were interested in discovering what potential existed for both sides to increase the level of participation in training in the future. What preferences did both parties have for the future?

Figure 25: Past and future Participation in training - Managers

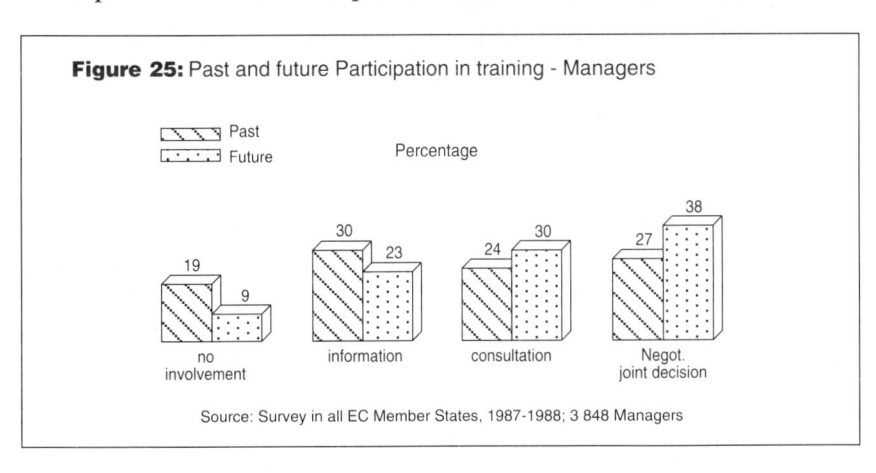

Past / Future · Percentage

	no involvement	information	consultation	Negot. joint decision
Past	19	30	30	38
Future	9	23	24	27

Source: Survey in all EC Member States, 1987-1988; 3 848 Managers

Figure 26: Past and future participation in training - Employee representatives

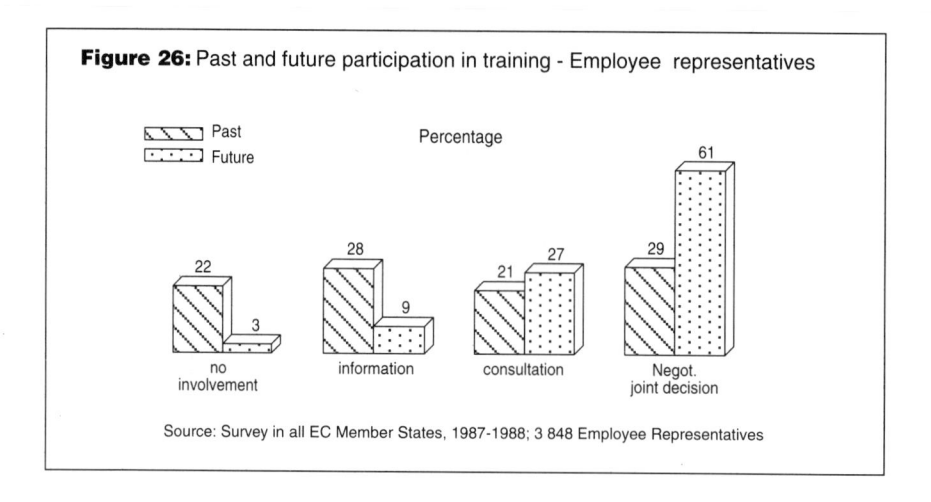

Source: Survey in all EC Member States, 1987-1988; 3 848 Employee Representatives

Figures 25 and 26 give a comparison of past and future participation levels in training at the overall European level. What is striking about the results is that there is a marked trend towards improved levels of participation in the future compared with the past. True, the expectations of employee representatives are much higher than those of managers as we would expect, but even if we take the responses of managers alone, the results show a much more positive trend in the future compared with the situation that existed at the time the interviews were conducted in 1987-88: the levels of "no involvement" have declined by about 10 percentage points; there is more consultation and negotiation and/or joint decision-making in training. It is clear that both sides are very much aware of not only the importance of training, but also the need for both sides to work closely together in the training area. The experience of the participation in training had been favourable.

Perhaps the most important point to note is that in every European Community country, there is a marked improvement in participation levels in the future compared with the past. Only in **Portugal** do we find that the improvement in participation shows only a marginal increase. If we take the views of managers as the best indicator of what future participation levels are likely to be, we notice that by comparing figures 27 and 28, the countries with the greatest improvement are **Denmark** and **Germany**. Those countries with the least improvement are **Portugal** and the **United Kingdom**.

In which countries is there the greatest level of disagreement between managers and employee representatives about the future? Here it is noticeable that **Belgium, France, Germany, Greece** and **Italy** show particular differences in opinion between the two sides.

Participation in Training

Figure 27: Future participation in training - Managers

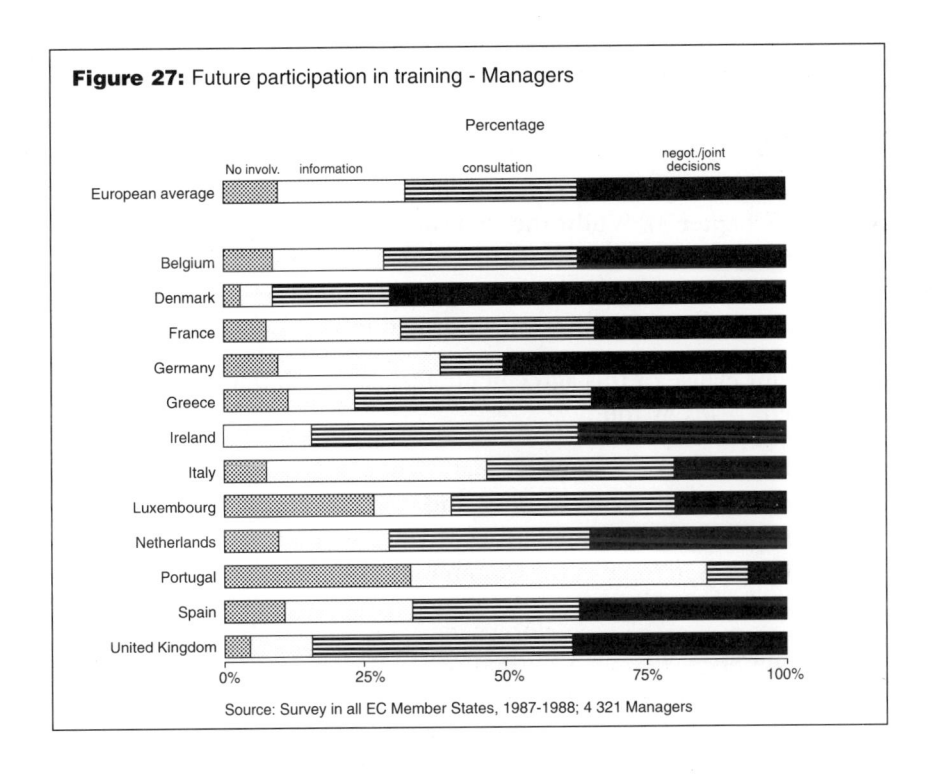

Source: Survey in all EC Member States, 1987-1988; 4 321 Managers

Figure 28: Future participation in training - Employee representatives

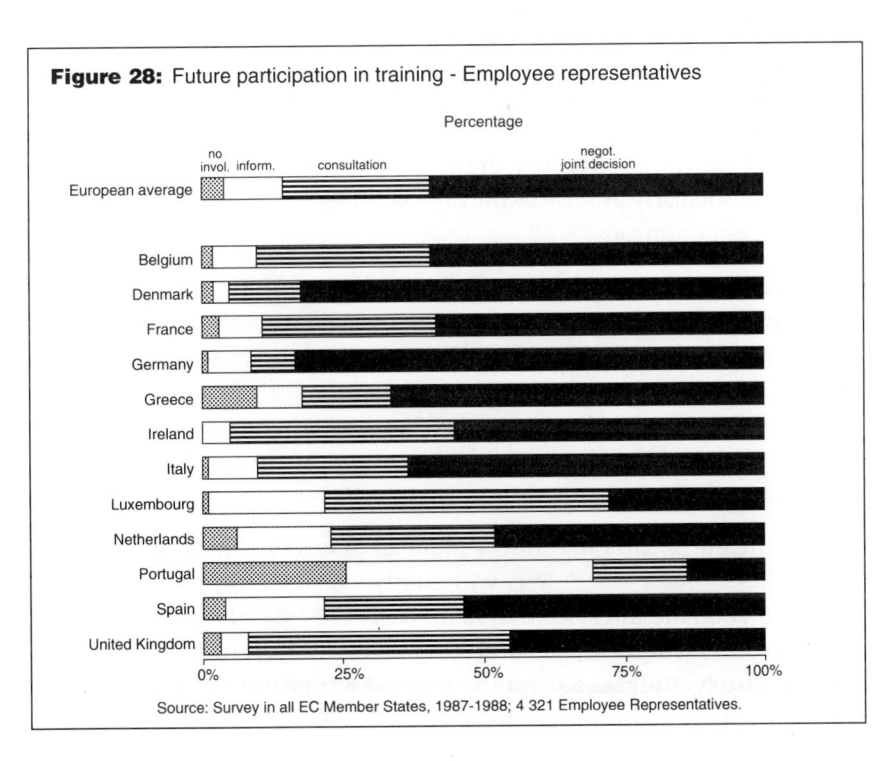

Source: Survey in all EC Member States, 1987-1988; 4 321 Employee Representatives.

6.8 Assessing the results in the light of the Val Duchesse Joint Opinion

One important benchmark against which our survey results can be measured is the Social Dialogue between the two sides of industry at European level, the ETUC on the one hand, and UNICE and CEEP on the other (see Chapter 3). Whilst the two sides to the Val Duchesse process have differing views as to how participation in training should be carried out, they were both signatories to two Joint Opinions — one on Training and Motivation and the other on Information and Consultation.

The relevant clause of this agreement, against which our results should be assessed, reads as follows:

> "Both sides take the view that, when technological change is introduced into a firm, workers and/or their representatives should be informed and consulted in accordance with the laws, agreements and practices in force in the Community countries. This consultation must be timely.

In this context:

(a) information means the action of providing the workers and/or their representatives, at the level concerned, with relevant details of such changes, so as to enlighten them as to the firm's choices and the implications for the firm's workforce:

(b) consultation of the workers and/or their representatives, at the level concerned, means the gathering of opinions and possible suggestions concerning the implications of such changes for the firm's workforce, more particularly as regards the effects on their employment and their working conditions".

We have interpreted this Joint Opinion to mean that employee representatives must be, at the minimum, consulted about training in new information technology. The mere provision of information about training does not, therefore, fall within the guide-lines laid down in the joint opinion. Of course levels of participation involving negotiation and/or joint decision-making about training are superior to the Val Duchesse provisions.

How do the European Community Member States rank in relation to the Val Duchesse guide-lines? We will firstly consider the situation that prevailed when the interviews took place in 1987-88. If we consider the responses of managers, we find that only in **Belgium, Denmark, Ireland, the Netherlands** and the **United Kingdom** did a majority of managers report that consultation or higher levels of involvement in training occurred. A middle-ranking group of countries which included **France, Germany, Greece, Italy** and **Spain** had between three and four managers out of every ten who claimed levels of participation that were within the Val Duchesse guide-lines. There was little compliance with Val Duchesse in the cases of **Luxembourg** and **Portugal.**

Employee representatives are slightly more sceptical about participation in training although there is a large measure of consensus between the two sides. Six countries — **Denmark, France, Germany, Greece, Ireland** and the **United Kingdom** have a majority of employee representatives who state that participation levels are consultative or at a higher level. There is a middle-ranking group of countries: **Belgium, Italy, Luxembourg**, and **Spain** while **Portugal** again appears at the bottom of the ranking order.

What is the likelihood that the Val Duchesse guide-lines will be observed in the future? There are three conclusions that can be made given the past experience of both sides. The most striking conclusion is that there is a significant movement of opinion on both sides towards greater levels of participation in the future in every European Community country. It is also evident that employee representatives have much greater expectations than their management counterparts in what they expect in the future; this, of course, is only to be expected. Thirdly, the pattern of preferences for the future, in terms of the ranking of countries, bears a close resemblance to that of the past.

Perhaps the best way of assessing the likelihood of compliance with Val Duchesse in the future is to consider the future preferences of managers. It is their attitudes, after all, which are important in determining the probable levels of participation that might emerge in the future.

At the aggregate European level around seven out of every ten managers indicate that training will be a matter for consultation or negotiation/joint decision-making in the future. In every country except **Portugal** there is a majority of managers who are likely to comply with Val Duchesse. There are seven countries where at least 70% of managers indicate that they will either be consulting employee representatives about training or making it a matter of negotiation/joint decision-making: **Belgium, Denmark, France, Greece, Ireland,** the **Netherlands** and the **United Kingdom.** Surprisingly, **Germany,** despite its renowned vocational training system, does not feature among these top seven countries. This can only be explained by noting the strong opposition that German managers have shown throughout the 1980s to any improvement in works council legislation in favour of employees. Of all the remaining countries, every one except **Portugal** has a majority of managers who indicate that they will be complying with the Val Duchesse guide-lines in the future.

6.9 Conclusions

It is generally recognised that training is one of the key priorities facing the European Community during the 1990s. The adoption of new information technology throughout the Community accentuates the need for much improved training of employees if the full potential of the Single Market is to be realised so that the Community can compete on favourable terms in world trade.

It is generally accepted that training provision is inadequate in many Community countries in relation to what is needed in order to enhance the productivity and competitiveness of European industry and that an increased effort is required. It seems from our data that there is no guarantee that participation in training will occur when new information technology is introduced into the workplace. Given that our sample covered only those firms which already had some form of institutionalised representation, one would expect that our data would indicate a much greater degree of partnership between managers and employee representatives in training matters than it does.

One feature of training in new information technology is that instead of relying on traditional specialist skills, it requires a greater emphasis on better mastery of general subjects, all-round skills and problem-solving experience. In other words, training in new information technology requires more of a 'modular' approach in order to take advantage of the greater degree of flexibility inherent in new information technology. All this requires a greater partnership between the two sides of industry, and participation by employee representatives in training is invaluable in utilising the full potential of new information technology.

What kind of picture emerges from our survey in respect of training? Perhaps the most important finding of our survey is that only half of the employee representatives were either consulted about training or had a stronger form of involvement in this important area. More ominously, given the importance that training occupies on the agenda of both managers and employees in the 1990s, it is particularly chilling to note that **one out of every five employee representatives have no involvement in training whatsoever**; this does not augur well for the future — particularly in light of the Commission's intention to propose an Instrument on access to training as part of the Social Charter.

Perhaps the most interesting part of our survey on the issue of training was the inter-country comparison of participation levels in training. There are marked differences in the level of co-operation that exists between both sides from one country to another. Sadly, there were many respondents of both sides who reported that there was no involvement at all by employee representatives in training at the time the interviews were conducted. In six of the 12 Member States less than half of employee representatives were consulted about training.

It is clear, however, that **managers in Europe are slowly recognising the importance of training and they increasingly see the need to work in a partnership with employee representatives**. The survey provides evidence that all countries of the Community are moving in different degrees towards a greater level of participation in the future. Moreover, **at European level, there is a high measure of agreement between the two sides of industry in relation to the Val Duchesse Social Dialogue.**

Given the sheer diversity of training provision throughout the Community from one country to another shown in this survey, Community policies to improve vocational training need to take account of these marked differences. Two of the key elements of Community policy in education and training — **producing a better qualified workforce and ensuring a greater equality of opportunity between regions** — might have a greater chance of success if our survey results showing greater levels of co-operation between the two sides of industry in the future prove to be correct.

NOTES
1. Held in Paris, 26-28 June 1991.
2. Commission of the European Communities, *Joint Opinions,* European Social Dialogue Documentary Series, 1991.
3. There are more than 3,000 such committees in the private sector.
4. For an excellent outline of the German vocational training system see Christel Lane, 'Vocational training, employment relations and new production concepts in Germany: some lessons for Britain, *Industrial Relations Journal,* Vol.21, No.4, Winter 1990.
5. Otto Jacobi and Walther Müller-Jentsch, "West Germany: continuity and structural change", in G. Baglioni and C. Crouch, *European Industrial Relations: the challenge of flexibility,* Sage, London, 1990, p. 148.

CHAPTER SEVEN

Participation in Health and Safety

7.1 Health and Safety at the European Community level

Among the different aspects of the social policies in the European Community, health and safety currently occupies a prominent place, although the interest of the European Community in health and safety at work is rather recent. In 1978, a decisive step was reached as the European Council adopted for the first time an action program in this area on the Commission's proposal. A second one was adopted in 1984. A number of directives came into force in that period.

However, it was not until the adoption in 1987 of the Single European Act that health and safety was given a great deal of serious attention, when this Act gave a new impulse to the development of Community legislation. Its Article 100A of the amended Treaty of Rome mentions "health" and "security" in the context of the proposals necessary for the establishment and the functioning of the internal market. More importantly, Article 118A states that "Member States shall pay particular attention to encouraging improvements, especially in the working environment, as regards the health and safety of workers, and shall set as their objectives the harmonisation of conditions in this area, while maintaining the improvements made." The measures to be taken by the EC Council in this area will be adopted by a system of "qualified majority" voting. Thus the adoption of directives in this field is made much easier than is the case in other matters of social policy for which unanimity in the EC Council is required (see Chapter 3).

Based on the above, the Council has issued a number of directives concerned with health and safety at the workplace. Among them, of prominent interest for this study, is the Framework Directive adopted in June 1989 on the introduction of measures to encourage improvements in the health and safety of workers at the workplace.

This Directive establishes the framework for the new European Community policy on workers' health and safety. Its scope is very wide, covering both private and public sectors and laying down a number of general obligations to be met by the employer. This directive is concerned with ensuring the provision of arrangements at the workplace by which standards can be achieved rather than by imposing the standards themselves. Therefore it includes principles such as the obligation to combat risks at source, to adapt technical progress and to give collective protective measures priority over individual protective measures. The Directive also deals with preventive services, first aid, fire-fighting and evacuation as well as the disclosure of information to workers and their consultation, participation and training. Employees' obligations include ensuring the correct use of machinery, the immediate reporting of hazards and appropriate co-operation with the employer and other workers. The Directive will come into effect in all the Member States by 31 December 1992 at the latest. Several other health and safety directives have since been adopted within the Framework Directive by the Council or are in preparation. Those adopted as of June 1991 deal with:

— minimum safety and health requirements for the workplace

— minimum safety and health requirements for the use of work equipment by workers at work

— minimum safety and health requirements for the use by workers of personal protective equipment

— minimum safety and health requirements for handling heavy loads

— minimum safety and health requirements for work with VDUs

The emphasis put on health and safety in the social policies of the EC further appears in the Charter of Fundamental Social Rights adopted by the European Council in December 1989. Article 22 of the Charter provides that "every worker must enjoy satisfactory health and safety conditions, more especially in his working environment..." Since then, the Commission has been actively pursuing the implementation of the Action Programme accompanying the Social Charter which contains a list of new initiatives dedicated to health protection and safety at the workplace. With these developments, the Community is now firmly established as the major force in the development of new health and safety legislation in the Member States.

The importance given to the adoption of minimum health and safety rules at Community level results from the proximity of the achievement of the single internal market and its increased competition. It is also in line with the growing awareness of the human and economic costs of occupational accidents and diseases. Indeed, economists have studied for a number of years the cost/benefit of prevention so as to determine the optimal level of security to be obtained in an enterprise.[1]

Prevention therefore appears to be more and more important and, by consequence, participation. For instance, a study of innovative workplace actions for health in seven EC countries recognises participation as being an essential prerequisite for successful workplace action.[2] This might explain why participation by workers in health and safety — or at least its principle — is the object of a consensus among the political authorities of the EC. Indeed, there is a willingness to involve the social partners at all levels in the elaboration and implementation of EC directives and Member States' policies as well as in enterprises.

The participation of workers has received special attention in the case of the introduction of new technologies: the Commission's programme concerning safety, hygiene and health at work as adopted by the European Council of December 1987 (88/C/28/02) contains a subject "information" in which the first point reads as follows: "In its joint opinion on information and consultation, the Val Duchesse Working Party stated:

> "When technological changes which imply major consequences for the workforce are introduced in the firm, workers and/or their representatives should be informed and consulted in accordance with the laws, agreements and practices in force in the Community countries." The Commission considers this objective to be particularly important where such practices have a potential impact on health and safety."

This position is reflected in the Framework Directive in which participative rights are clearly established. According to article 1.2, the directive "contains general principles concerning the prevention of occupational risk and accident factors, the **information**, **consultation**, **balanced participation** in accordance with national laws and/or practices and training of workers and their representatives, as well as general guidelines for the implementation of the said principles."

Its most relevant provisions on worker participation are to be found in Articles 10 and 11. In Article 10, it states that "the employer shall take appropriate measures so that workers and/or their representatives in the undertaking and/or establishment receive... all the necessary **information** concerning:

(a) the safety and health risks and protective and preventive measures and activities in respect of both the undertaking in general and each type of work-station and/or job..."

Article 11 provides that "employers shall consult workers and/or their representatives and allow them to take part in discussions on all questions relating to safety and health protection at work. This pre-supposes:

— the **consultation** of workers;

— the right of workers and/or their representatives to make **proposals**;

— **balanced participation** in accordance with national laws and/or practices."

Special mention of new technologies is made in Article 6 dealing with general obligations on employers. They must "ensure that the planning and introduction of new technologies are the subject of consultation with the workers and/or their representatives, as regards the consequences of the choice of equipment, the working conditions and the working environment for the safety and health of workers". (Art. 6.3.c). Compared to other aspects of worker participation in the introduction of new technologies, health and safety has received particular attention from the political authorities of the EC.

This European context constitutes the background against which we can examine the survey results concerning involvement of employee representatives in the improvement of the working environment, health and safety.

7.2 Health and Safety within the Member States

Another element to be taken into account concerning participation in health and safety is the legislation existing in the EC Member States regarding workers' representatives rights to information and consultation with their employers on matters affecting their health and safety at work. Nearly all the countries surveyed have legislative provisions in this area. Some, like **Belgium**, have had them for several decades while in other countries such as **Greece** they are of much more recent origin. In most EC countries the main relevant legislation now in application was enacted in the 1970s (see TABLE 3 on p. 116). However, legislation providing for the involvement of worker representatives in health and safety had existed prior to the present legislative provisions — in **Germany** (1952), **France** (1974/1979), **Ireland** (1955), **Luxembourg** (1925), the **Netherlands** (1950) and **Spain** (1944). Much importance has thus been given to employee involvement in health and safety and we would therefore expect a high degree of worker participation in most of the 12 Member States.

If we consider the type of institutional approach which has been adopted to achieve worker representation in health and safety at the workplace, there are several different approaches which can be identified throughout the European Community.[3]

In one group of countries, rights of information and consultation on health and safety are granted exclusively or mainly to representatives in mandatory specialised bodies of joint regulation within establishments.

In **Belgium**, rights are granted by law to workers' representatives on health and safety committees which are obligatory in firms with more than 50 employees. In **France**, health, safety and improvement of working conditions committees are compulsory in firms with 50 or more employees. Rights are very explicit and detailed in these two countries. In **Denmark**, special rights are granted by law to safety representatives

— elected by employees in all undertakings with more than 10 employees — and also to workers' representatives on safety committees which are obligatory in all firms with more than 20 employees. Compared to the arrangements adopted in most other Member States, the Danish health and safety system is characteristic in its comprehensiveness and in the extent to which the principle of co-operation is embodied in the institutions provided by the law for the purpose of worker involvement. In **Greece**, special rights in this field are granted to representatives on health and safety at work committees elected in undertakings employing more than 20 workers. In **Ireland**, rights are granted to workers' representatives on safety committees — obligatory in all workplaces covered by the 1980 Act (i.e in manufacturing industry, construction and those places which fall within the definition of a factory) with 20 or more employees — and to safety representatives — elected by employees in firms with less than 20 workers. Safety committees have been established by collective agreement in many industries not covered by the Act. It is estimated that more or less 25% of the workforce is covered. Under the present system of statutory law, the rights of workers or their representatives to be involved in health and safety seem very limited, regarding both information and consultation. In **Spain**, special rights are granted to representatives on health and safety committees which must be set up in firms with more than 100 employees and in firms with less employees in cases where the Ministry of Labour and Social Security considers that the health and safety risks render it necessary.

In addition, certain rights in this area are also granted to representatives on the works council in **France**, **Spain** and **Greece** (where the relevant legislation dates back from 1988). It should be noted that in several countries of this first group, no statutory provision exists concerning the establishment of works councils: it is the case in **Denmark** and **Ireland**. This privileged status given to health and safety matters indicates their importance.

In another group of countries, all rights in the area of health and safety are conferred on workers' representatives on works councils. In **Germany**, special rights are granted to representatives on works councils elected by and from employees in all undertakings with five or more workers. Also covered are employee safety representatives who are employees appointed by the employer after consultation with the works council in all firms with more than 20 employees. In firms where there is a company doctor or safety expert, or where there are more than three employee safety representatives,a work safety committee must be set up, on which two works councillors sit. Works councillors have the right to co-determination in all health and safety matters. In **Luxembourg**, special rights — including co-determination in health and safety matters — are granted to representatives on the works council, elected by employees in all enterprises employing 15 or more workers. Specifically, rights are granted to a safety delegate designated by the works council from their own members or from the workforce as a whole. In the **Netherlands**, special rights are granted to workers' representatives on the works council — which must be set up in all companies with at least 35 employees. In larger undertakings the works council can delegate some of their rights in this area to health, safety and welfare committees, a majority of whose members are works councillors and the rest other employees. Works councillors have information and consultation rights and they must give their agreement when the employer wants to enforce, change or suppress health and security rules.

In a third group of countries, there are no mandatory regulations in this area. The legislation supports the institutions of worker representation at the workplace by providing them with rights they may exercise if they choose.

In **Italy**, the Workers' Statute confers rights in the area of health and safety — among others — on the "most representative unions" at enterprise level. It applies to operating units of industrial/commercial firms employing more than 15 workers, to agricultural enterprises with over five workers and to public authorities. It allows employees from the "most representative unions" — though this phrase was left undefined — to form enterprise-level union organisations (known collectively as *"rappresentanze sindicali aziendali"* or RSAs). The law does not regulate the organisation or operation of RSAs but the most widespread form is the works council *(consiglio de fabbrica)*. Although the Workers' Statute provides rather extensive rights, such as control of the implementation of the legislation, there is no statutory unambiguous right to information and consultation for employee representatives but such a right is often included in collective agreements. There is much debate in the legal literature concerning the interpretation of the law on this point and the most usual answer concludes to the implicit existence of a right of information: without it, employees' representatives could not exercise their other rights.

The **Portuguese** system of worker involvement in health and safety is also rather voluntary in its character. The establishment of safety committees and safety delegates depends in most sectors on what is agreed upon by the social partners in collective agreements. The establishment of workers' committees and the appointment of trade union delegates is optional and conditional upon the initiatives of employees and trade unions. Legal rights of workers and their representatives in health and safety matters are laid down in the law, but only to a limited extent.

In the **United Kingdom**, special rights, including information and consultation, are granted by law to safety representatives appointed by trade unions from the workforce in workplaces where the trade union is recognised. Rights are also granted to workers' representatives on safety committees which must be set up on the request of two safety representatives or can be formed through collective agreement or employer initiative. Other things being equal, one might expect that in countries where there are specialised worker representation institutions and where safety and health matters form a subject for information and consultation apart by statute from other employment related matters, more attention is paid to these questions and therefore conditions are more favourable for participation by employers' representatives.

From this point of view, the situation in the different countries is summarised in TABLE 3 which also shows when the presently applicable

Table 3
Institutional Arrangements and Information and Consultation Rights in Health and Safety

Rights are granted to workers' representatives on: Content of information & consultation rights	No statutory unambiguous right to information & consultation in health & safety matters	General & unspecific right in health & safety matters	Specific information & consultation on certain issues as well
Health and safety committees		*Denmark* (1975)	*Belgium* (1952) *France* (1982) *Greece* (1985/88) *Ireland* (1980) *Spain* (1971/74/80)
Works councils		*Germany* (co-determination) (1972/73) *Luxembourg* (co-determination) (1979)	*Netherlands* (co-determination) (1982)
Bodies to be triggered at the workplace	*Italy* (1970)	*Portugal* (1975/79)	*United Kingdom* (1974/77)

The figures between parentheses indicate the years when the presently applying relevant legislation was adopted.

Table 4
Workers' Representatives' Rights in Health and Safety

	Favourable	Neutral	Unfavourable
Institutional arrangement + depth of information & consultation rights	Belgium Denmark France Germany Greece Luxembourg Netherlands	Ireland Spain	United Kingdom Italy Portugal
Relations with the inspectorate*	Belgium Netherlands Germany Italy Ireland Luxembourg Spain United Kingdom	Denmark France Greece	Portugal
Suspending work**	Denmark Spain Ireland Netherlands Luxembourg	Greece	Belgium France Germany Italy Portugal United Kingdom

* Favourable = previous consultation by labour inspectorate of workers representatives before it decides on certain measures and/or right to call the inspectorate
Neutral = right to meet an inspector visiting the establishment or to accompany him on his inspection tour
** Favourable = in case of danger, right to halt the work or to call the inspectorate
Neutral = right to demand action from employers.

legislation was adopted.) Taking this into account, as well as the depth of rights of information, consultation and co-determination, we would expect participation in health and safety to be highest in **Belgium** and **Denmark**. Among the countries without specialised bodies, it should be highest in **Germany**, **Netherlands** and **Luxembourg** and it should be at its lowest in **Portugal**.

7.3 Survey results[4]

One of the questions asked in the survey concerned a number of issues on which employee representatives could be involved in decisions about new technology. For each of these issues — in the present case "improving the working environment, health and safety" — the exact wording was: "When this new technology was introduced, how were employee representatives involved?"

Let us first emphasise that, as appeared in previous stages of this research, there is definitely more participation in problems of relevance to employees than in matters of primary importance to managers. And, among the former, health and safety appears as the workforce concern for which both managers and employee representatives report the highest levels of participation, in the twelve countries taken together: according to one out of three respondents, negotiation and/or full co-determination was used when introducing new technologies.

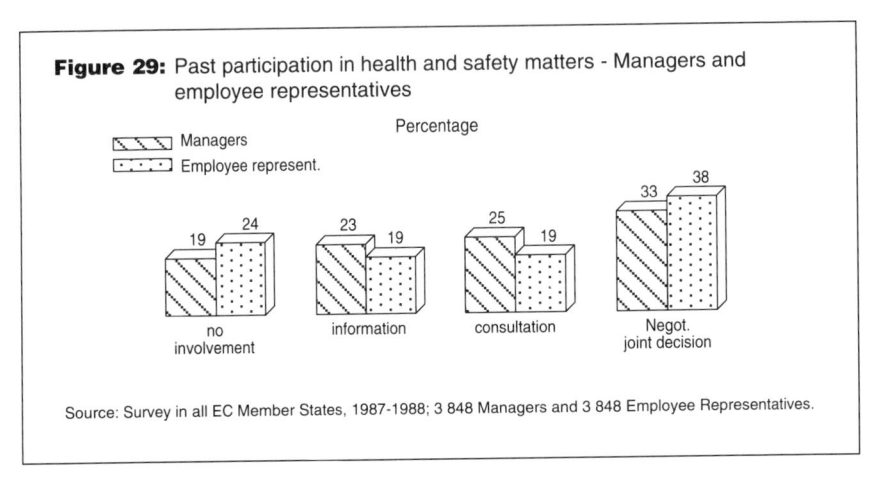

Figure 29: Past participation in health and safety matters - Managers and employee representatives

Percentage

Managers
Employee represent.

	no involvement	information	consultation	Negot. joint decision
Managers	19	23	25	33
Employee represent.	24	19	19	38

Source: Survey in all EC Member States, 1987-1988; 3 848 Managers and 3 848 Employee Representatives.

Yet, if we measure participation by the Framework Directive's yardstick which requires information and consultation, we find it is not met in around 60% of the firms.

Managers' answers do not differ much from employee representatives' answers: the latter report a somewhat higher frequency of no involvement and of co-determination and less information and consultation; but the difference between the two categories of respondents does not go beyond 6 percentage points.

Let us now turn to the comparison between countries. The level of employee participation in health and safety at work differs significantly among the twelve Member States of the European Community, one extreme being represented by the Federal Republic of **Germany** and the other by **Portugal**. In **Germany,** according to managers, employee representatives enjoy full co-determination in 63% of the firms and receive no information in only 8% of the firms. The situation in **Portugal** indicates the opposite: 67% of the firms do not impart information on health and safety when introducing new technologies while only 3% of them provide negotiation or joint decision-making. Consultation and higher forms of participation represent less than 5%.

PARTICIPATION IN HEALTH AND SAFETY

Figure 30: Past participation in health and safety matters - Managers

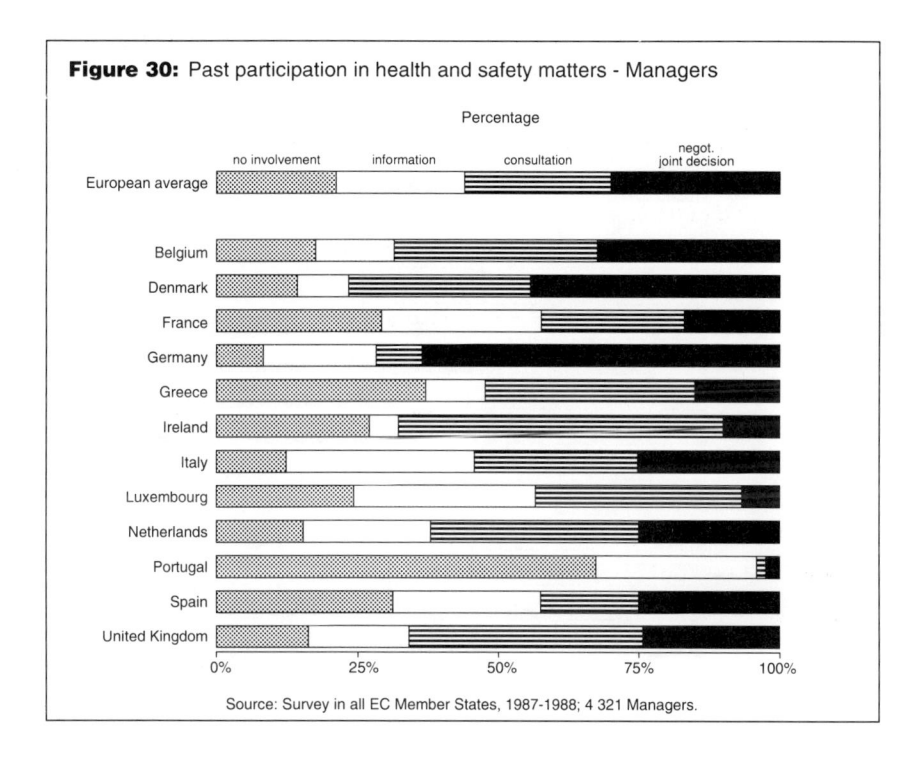

[Bar chart showing percentages of no involvement, information, consultation, and negotiation/joint decision across European average, Belgium, Denmark, France, Germany, Greece, Ireland, Italy, Luxembourg, Netherlands, Portugal, Spain, and United Kingdom.]

Source: Survey in all EC Member States, 1987-1988; 4 321 Managers.

Comparatively, **Denmark** and **Belgium** fare rather well in this area of health and safety: 77% and 69 % of managers, respectively, report consultation and higher forms of participation in these two countries where the level of co-determination is also above average.

Managers in Ireland, United Kingdom and Netherlands also report at least consultation for around two thirds of them but the level of co-determination is lower in these countries.

Luxembourg and the southern countries of Europe, **Italy, Greece, France** and **Spain** appear to show a comparative deficit in this field: in **Spain,** more than two managers out of three do not report more than information in this important area.

If we compare these results with the employee representatives' answers, they convey a rather similar picture although in **Belgium, Ireland** and **Netherlands,** the employee representatives have a less positive view of participation than their manager counterparts.

These results for the different countries demonstrate the absence of a straightforward relationship between the existence of legislation favourable to information and consultation in health and safety and *effective* participation. It seems that despite the fact that there are

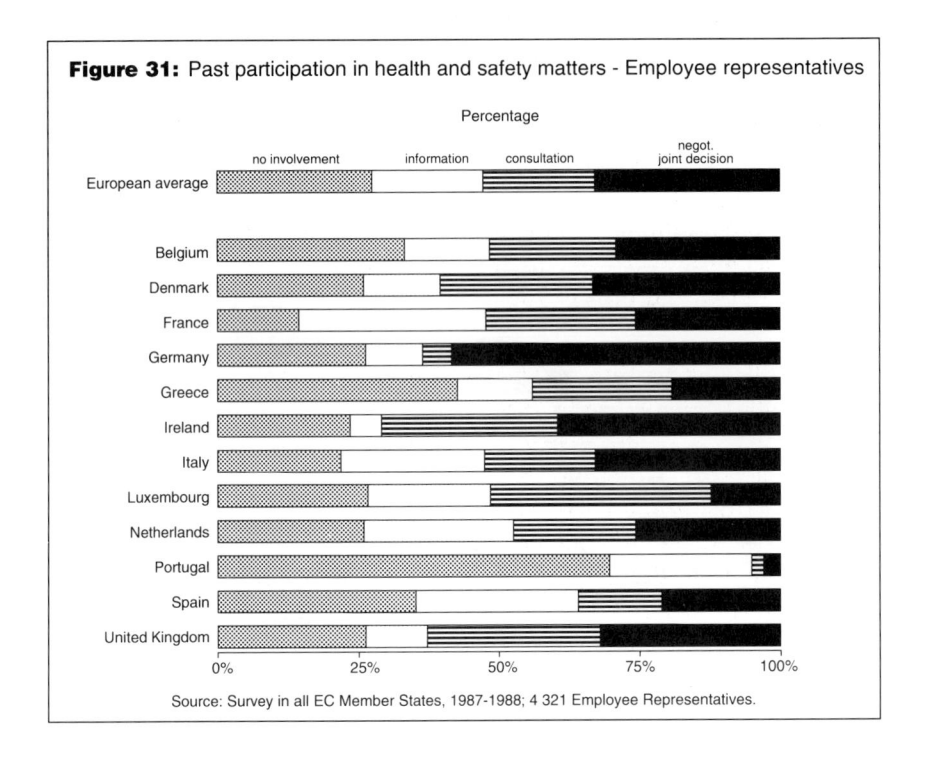

Figure 31: Past participation in health and safety matters - Employee representatives

Source: Survey in all EC Member States, 1987-1988; 4 321 Employee Representatives.

legislative provisions pertaining to participation in this area in all the countries surveyed, they are not systematically observed. This is not really surprising as previous work pertaining to works councils has shown that "there is some evidence that practice diverges — perhaps considerably — from statutory requirement".[5]

Nevertheless, statutory regulations can be linked to the comparative level of participation in **Germany**, where works councillors have the right to co-determination and may even take the initiative in all health and safety matters. It is also the case for **Denmark**, for **Belgium** and for **Portugal**.

At the other hand, if the existence of health and safety committees with information and consultation rights were a determinant factor, we would certainly expect **France** to display higher levels of participation.

We would also expect better scores in **Greece**. However, the legislation there is quite recent: workers' representatives' health and safety rights are contained in a law of 1985 on occupational health and safety and in a law of 1988 on works councils.

There is no unequivocal relationship between legislation and participation if we examine other aspects of the rights of workers' representatives in health and safety such as the general and unspecified right to information and consultation versus entitlement to certain specific types of information and consultation on certain issues as well (see Table 3) or relations between workers' representatives and inspections or even the right to halt the work in an undertaking on the grounds of danger to health and safety. Yet, this result should be interpreted with caution and as no more as a hypothesis: it is difficult to know from the texts — and still more so from a summary of them — the exact implications of legislative provisions. Moreover, we do not know whether other measures of participation would lead to the same results.

Given our data, legislation should be seen as merely one favourable factor among others which is conducive to participation. The *effectiveness* of the statutory provisions is probably another matter and, as has already been suggested[6] depends on factors such as strong trade union workplace organisation. The results obtained for the **United Kingdom** (where joint consultative machinery is common) seem in line with this explanation as do the poor results for the southern European countries. Indeed, on the whole, the variables defined in the explanatory frame of reference set out in chapter 4 seem to have more explanatory power than the mere existence or otherwise of legislation providing for worker representatives' involvement in health and safety matters.

What kind of picture do we find in the future?

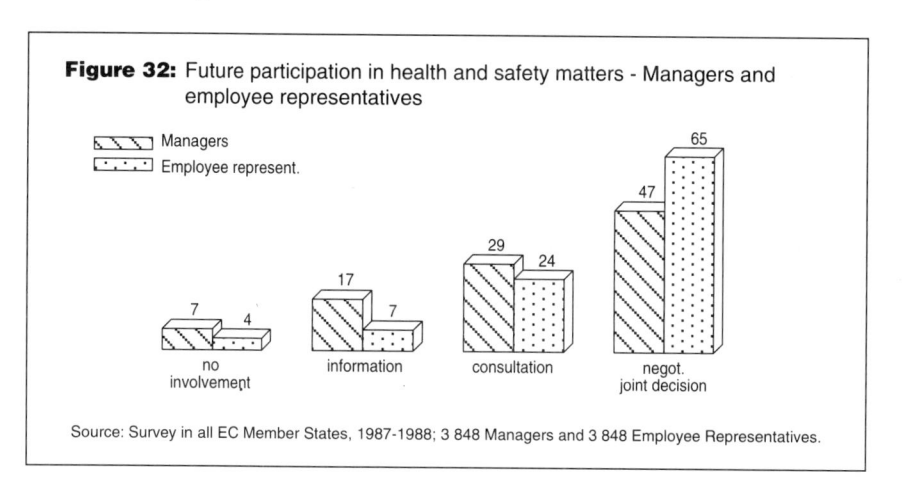

Figure 32: Future participation in health and safety matters - Managers and employee representatives

Managers
Employee represent.

	no involvement	information	consultation	negot. joint decision
Managers	7	17	29	47
Employee represent.	4	7	24	65

Source: Survey in all EC Member States, 1987-1988; 3 848 Managers and 3 848 Employee Representatives.

When we look at expectations concerning the future, both sides of industry anticipate more consultation and more co-determination but the latter form of participation is more desired by employees: 65% of them against 47% of managers.

WORKPLACE INVOLVEMENT IN TECHNOLOGICAL INNOVATION IN THE EC

What are the expectations of the social partners of the different European countries about participation in the area of health and safety?

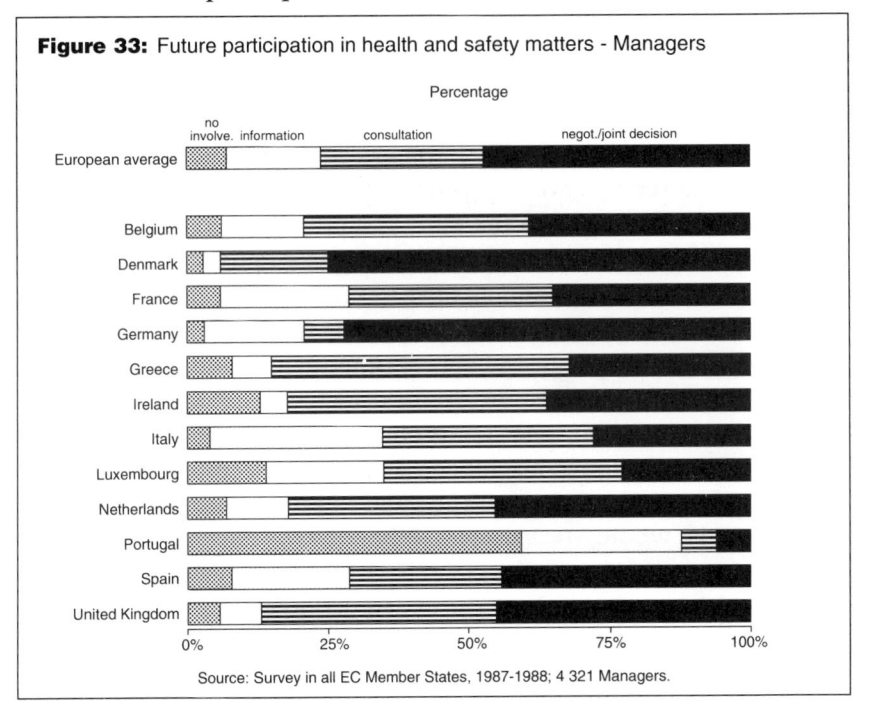

Figure 33: Future participation in health and safety matters - Managers

Source: Survey in all EC Member States, 1987-1988; 4 321 Managers.

The European average figures for managers and employee representatives conceal major national differences. In **Germany**, where the actual level of co-determination is high, the shift between past and future is rather small for the managers: from 63 to 70%. It is striking to see that the level obtained in **Denmark** is higher: if less than one out of two managers judged that co-determination was taking place in the past for health and safety, this proportion increases to three out of four in the future. In any case, the percentage of managers desiring intensive employee participation in these two countries is greater than the percentage of employee representatives expressing the same desire throughout all of Europe.

In the **United Kingdom**, the **Netherlands** and **Spain**, more than 40% of managers support full involvement. **Portugal** once again takes up the rear, with more than 60% of managers refusing any form of information, even in the future; but there is also little support to be found among company managements in **Italy** and **Luxembourg**.

It should be noticed that, in **Greece**, **Spain** and **France**, around 30% more managers favour consultation and higher forms of participation in the future. This might be linked to the promises contained in the legislation in this area which, at the time when the survey was conducted was still quite recent or, in **Greece**, not yet enacted.

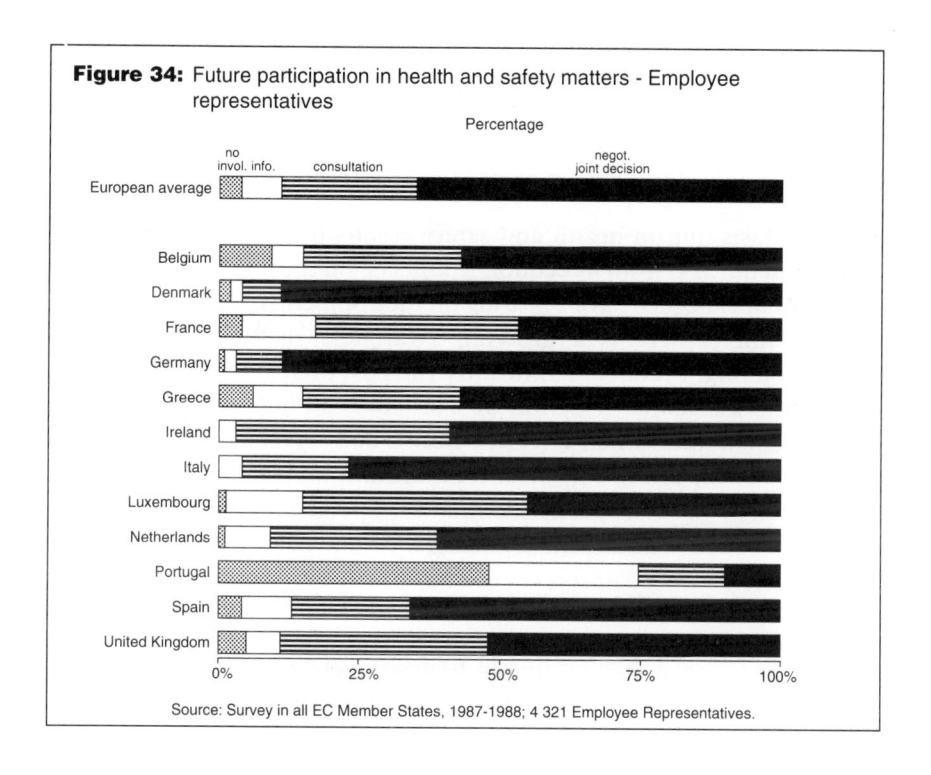

Figure 34: Future participation in health and safety matters - Employee representatives

Source: Survey in all EC Member States, 1987-1988; 4 321 Employee Representatives.

Similar differences are to be found among employee representatives. 75% to 85% of **German**, **Danish** and also, with some disparity, **Italian** employee representatives demand full co-determination or negotiation with equal entitlement on the protection of health and safety. If a wide difference exists most of the time between the preferences of both sides of industry for the future, it appears especially wide in this country: one out of four managers desire equality of rights in this area compared with three out of four employee representatives.

Well over half the employee representatives in **Ireland**, **Belgium**, the **Netherlands**, **Spain** and **Greece** support this position. The level of support in the **United Kingdom**, **France** and **Luxembourg** is 40% to 50%. Only 10% of employee representatives in **Portugal** take a favourable view of co-determination and almost half of employee representatives reject information in this area in the future.

7.4 Conclusions

While legislation in this area already pre-existed in all the Member States, health and safety at work gained gradually in importance in the social policies of the European Community and now occupies a prominent place on the European political agenda. This appears clearly in article 118A of the Single European Act, in the Charter of Fundamental Social Rights as well as in the Action Programmes and the Directives which have been issued. Among the latter, the Framework Directive on the introduction of measures to encourage improvements in the health and safety of workers at the workplace is of particular interest for this survey.

The emphasis put on health and safety results not only from its social dimension but also from its economic aspects: on the one hand, minimum standards in existence have a bearing on the competitiveness of firms and, on the other, there is a growing awareness that good systematic prevention might reduce significantly the cost of accidents and occupational diseases. As participation of workers is an important element in the implementation of an effective health and safety policy in firms, a consensus exists around it, or at least around its principle, among the authorities of the European Community. Specifically, the Framework Directive requires "information, consultation and balanced participation" in this area.

Among the issues of prominent interest to workers, health and safety appears to foster a comparatively higher level of participation in the process of introducing new technologies. Nevertheless, wide differences exist between the countries of the European Community, both in practice and in future expectations.

According to managements' responses, there are only two countries where at least 70% of the firms respect the consultation objective of the Framework Directive on health and safety — **Germany** and **Denmark**. **Belgium**, **Ireland**, **United Kingdom** and **Netherlands** are still above 60%. **Luxembourg** and the Southern countries of Europe, including **France**, are in a worse position, especially **Portugal**, where consultation and higher forms of participation are not even practised in 5% of the firms.

The relationship between these results and employee representatives' rights of information and consultation in the national legislation is not straightforward. On the whole, the explanatory factors set out in Chapter 4 such as technological objectives, management style, the bargaining power of organised labour, regulation and the degree of centralisation of the industrial relations system seem to explain varying degrees of participation in health and safety matters in our survey results.

The deficit between employee participation in practice and the Framework Directive's minimum requirement will markedly reduce within a foreseeable period of time. Only 24% of managers and 11% of employee representatives in Europe are sceptically opposed to a satisfactory practice of consultation within the near future. Moreover, we can expect employee representatives to push in that direction: they are much more in favour of co-determination than their management counterparts as over two thirds of them seek co-determination in these matters. So, the implementation of the EC objectives of information, consultation and participation concerning the health and safety aspects of new technologies seems to have a bright future.

NOTES
1. For a good synthesis of work in this area, see B. Brody, Y. Letourneau et A. Poirier, "Le coût des accidents de travail: état des connaissances", *Relations industrielles,* vol. 45, no. 1, hiver 1990, pp. 94-117.
2. R. WYNNE, *Innovative Workplace Actions for Health: An Overview of the Situation in Seven EC Countries,* Working Paper EF/WP/90/35/EN, Dublin, European Foundation for the Improvement of Living and Working Conditions.
3. The following is based on J.K.M. Gevers, *Health and safety protection in industry: Participation and information of employers and workers,* CEE, Directorate General Employment, Social Affairs and Education, Report EUR 11314EN, 1987.
"The rights of workers' representatives: health and safety", *European Industrial Relations Review,* no.183, April 1989;
Dr. Walters, *Worker Participation in Health and Safety: A European Comparison,* London, Institute of Employment Rights, 1990;
La dimension sociale du marché intérieur, troisième partie: la représentation des travailleurs dans les entreprises d'Europe occidentale, Info 32, Institut syndical européen, Bruxelles, 1990;
M. Gold and M. Hall, *Legal Regulation and the Practice of Employee Participation in the European Community,* Working Paper EF/WP/90/41/EN, Dublin, European Foundation for the Improvement of Living and Working Conditions.
4. The following analysis owes much to Dr. H.Krieger, *Participation of employee representatives in the protection of health and safety of workers in Europe,* Mitbestinnmung, No.11, 1990, where the same data are examined.
5. M. GOLD and M. HALL, *op. cit.,* p. 37.
6. Dr. WALTERS, *op. cit.,* p. 21.

CHAPTER EIGHT

Participation in Work Organisation

8.1 Introduction

One of the major concerns of employee representatives about the introduction of new information technology into their organisations is the danger that it will be accompanied by forms of job design and a division of labour which are based on "Taylorist" principles of "scientific management". The principles of scientific management were first espoused by Frederick Taylor at the turn of the century and such principles have formed the core of mass production techniques, and have spread to batch production as well as office and service work in many work organisations. They involve features such as external job control; job division and fragmentation; technological control (e.g. through the assembly line); repetitive work; deskilling; work measurement and time and motion study; individualised control (e.g. incentive payment systems); and finally, very little scope for social interaction by workers.

The negative consequences of such forms of work organisation are well-known: it frequently leads to poor worker motivation and low morale as well as declining worker effort. Employees working under such systems have boring, repetitive and routinised jobs; they are denied responsibility; they lose interest in their work; they are closely supervised and have little or no commitment to the goals of the enterprise. Whilst it can be said to be "efficient" and "cost effective", it is often argued that the disadvantages in worker morale outweigh the economic advantages for management. Yet even though Tayloristic principles of job design generated a great deal of concern, manifested in such trends as the "human relations", the "quality of working life" and "work humanisation" schools of the 1960s and 1970s, all seeking in their own way to mitigate the adverse effects of Taylorism, Tayloristic systems are still widespread throughout industry and commerce.

The Quality of Work Life (QWL) programmes of the 1960s and 1970s were essentially concerned with introducing measures such as job rotation, job enlargement, job enrichment, the relaxation of work and time rules and the introduction of team and group work. The most famous examples of these were in Sweden — at the Volvo plant in Kalmar (and latterly at Uddevalla) and the Saab plant at Trollhatten; both of these schemes sought to eliminate the alienating aspects of assembly line work. Similar schemes of group working have recently been introduced at the GM plant in Austria and Germany. Restructuring of job classifications which seek to upgrade the skills of production workers have been introduced into the Wolfsburg Volkswagen plant.

The QWL programmes which were introduced in the 1970s were largely designed to ameliorate the worst aspects of Taylorism; they were not motivated by the need to reorganise production in order to take full advantage of the potential possibilities of computerisation. It is now clear that new information technology, with its wide-ranging potential for radically altering the way work is carried out — both on the shopfloor and in the office environment — offers new opportunities for both sides of industry to redesign the way work is organised. The pressures of competition emanating from more varied and specialised world markets allow more flexible and adaptable methods of work organisation.

These new forms of work organisation are designed to optimise the flexibility that derives from programmable computerised systems. It is clearly in the interests of trade unions to draw on the experience gathered from earlier QWL programmes to ensure that this flexibility is based on collective, democratic structures, rather than on atomised, individualised anti-union structures. New information technology offers management more choice in the way work is organised; it is not deterministic. Daniel's study in the UK shows that workers are only likely to oppose technical change when it is accompanied by major changes in work organisation. It is therefore to be expected that employee representatives in European enterprises will seek to avoid substituting the inflexibility of Tayloristic forms of job design for forms of flexibility which seek to deskill workers. Participation in work organisation can therefore have a major role to play in ensuring that the introduction of new information technology is in the interests of workers.

There is much variation across the European Community in the premium that is placed by trade unions in different Member States on the importance of good work organisation. In some countries, notably in **Denmark, Germany** and to a lesser extent the **Netherlands**, unions have traditionally taken a strong interest in job design.

In **Denmark**, this has been manifested through the auspices of Co-operation Committees which provide for a separate addendum to the 1970 central agreement between the LO and DA on new technology. Under this agreement, either party can call in a 'special expert from the undertaking', and if both parties agree, to call on 'other experts'. Whilst Danish practice in issues concerned with work organisation is considerably less wide-ranging than is the case in Sweden, it is still true to say that the attitudes of the Danish LO on such issues falls within the Scandinavian tradition.

Trade unions in **Germany** have sought to open up works council activities to influence and support from external bodies and to build on the experience the unions gained in the *Humanisierung der Arbeit:* HdA and the *Produktions und Fertigungstechnik* programmes of the late 1960s and the early 1970s. In addition to this there is provision for State financial support for QWL initiatives at workplace level managed jointly by employers and works councils in a corporatist or co-decision relationship. The DGB has also established Technology Advice Centres *(Technologieberatungsstellen),* which provide support and advice to works councils on new technology issues, of which issues concerned with work organisation occupy a prominent part.

We would expect our survey results to show a high degree of participation in work organisation in both these two countries. Both countries have well-resourced trade union movements, well-trained union officials, well-established traditions of concern for working life initiatives from both sides, management traditions that are not openly averse to issues of work organisation and some degree of governmental support. In contrast, we would expect countries with trade union movements which are either divided along religious or political lines and/or which are poorly resourced to pay less attention to QWL issues; **Portugal** and **Greece** are examples. Trade unions in some Member States which, although well-established and having some degree of bargaining power, have long concerned themselves with pay and job security issues and have placed little emphasis on job design issues; the **United Kingdom** is one such example.

In summary, work organisation and quality of working life issues have only been given prominence in the countries of northern Europe. Trade unions in **Denmark, Germany,** the **Netherlands** and **Belgium** have all been concerned to a greater or lesser extent with the way new technology shapes jobs, and have recognised the importance of being involved in the design stage of new information technology. In contrast, trade unions in countries in the southern part of the Community, e.g. **France, Italy, Portugal, Spain** and **Greece** attach little importance to this aspect of new information technology. We would expect this contrast to be reflected in the survey results.

8.2 Participation at European level

Figure 35 illustrates the levels of participation in existence at the European level at the time the interviews were carried out according to managers, together with their preferences for the future. Managers reported the lowest level of no involvement for work organisation among the workforce concerns investigated — information only being the most frequent type of involvement reported. Around three out of four interviewees stated that employee representatives were at least informed about changing work organisation and the division of labour. Around

PARTICIPATION IN WORK ORGANISATION

Figure 35: Past participation in work organisation - Managers and employee representatives

Managers
Employee represent.

no involvement: 18%, 23%
information: 33%, 31%
consultation: 23%, 19%
negot. joint decision: 26%, 27%

Source: Survey in all EC Member States, 1987-1988; 3 848 Managers and 3 848 Employee Representatives.

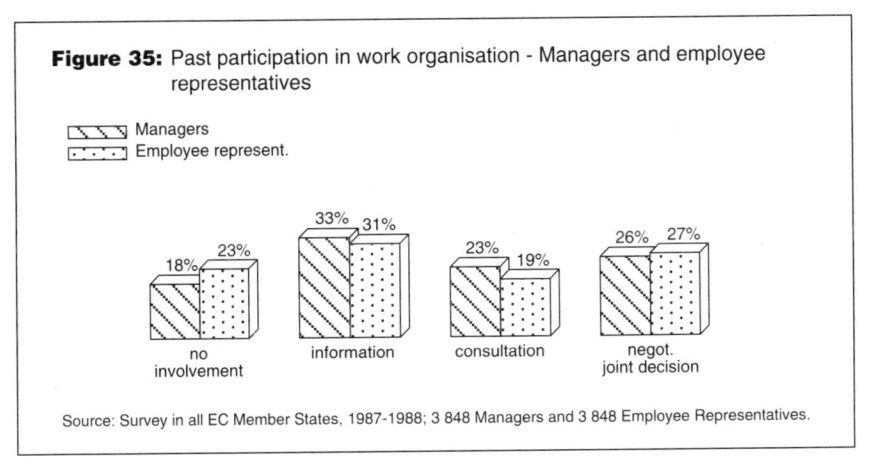

a quarter of the respondents estimated that some form of co-determination was the type of participation used for work organisation.

8.3 Participation at Member State level

What are the differences appearing between the twelve countries of the European Community? If we take as a reference point the European average of 51% for no involvement and information, the managers of the following countries report lower percentages for these two levels: **Ireland, Denmark, Belgium, United Kingdom, Netherlands** and **Germany**. Among those, it is in **Germany, Netherlands** and the **United Kingdom** that it is least frequent not to involve employee representatives: in these three countries, hardly more than one out of ten managers report such a situation.

On the whole, the answers of the employee representatives are comparable. Yet, in several countries employee representative respondents have a more pessimistic view of participation than managers. Generally, these six countries represent the top half for participation in the introduction of new information technology.

We have already explained how the different countries fare with respect to the five main variables identified in chapter 4 where an explanatory framework of country-specific factors which influence participation levels was developed. Our results can be largely explained here by referring to technological objectives of management, management style, the bargaining power of unions, forms of regulation and the degree of centralisation of the industrial relations system.

Figure 36: Past participation in work organisation - Managers

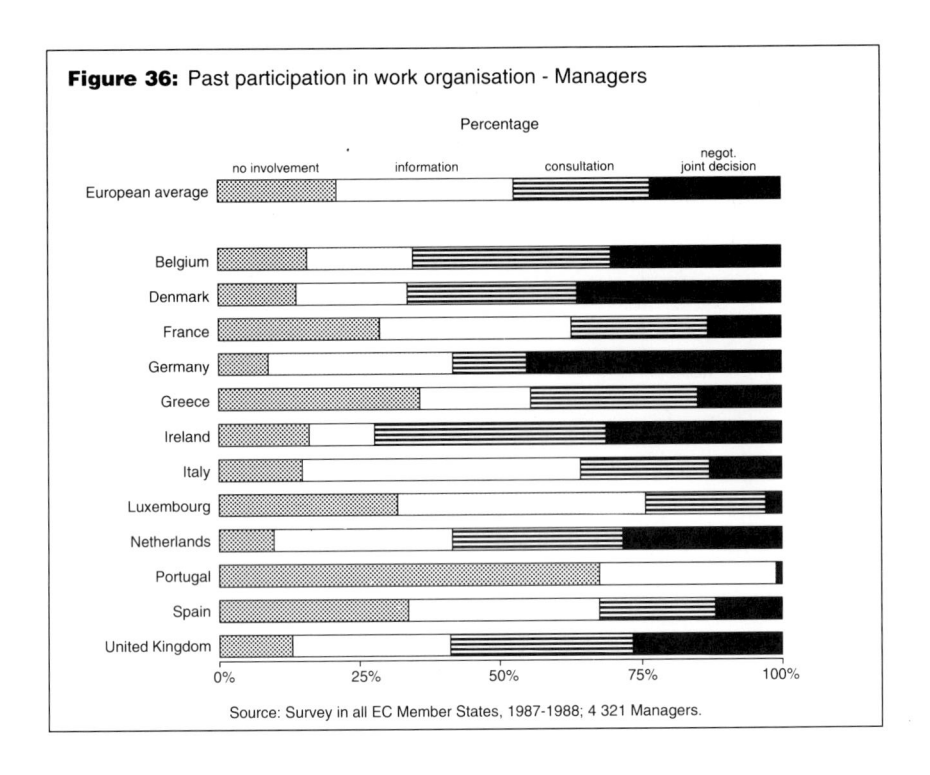

Source: Survey in all EC Member States, 1987-1988; 4 321 Managers.

Figure 37: Past participation in work organisation - Employee representatives

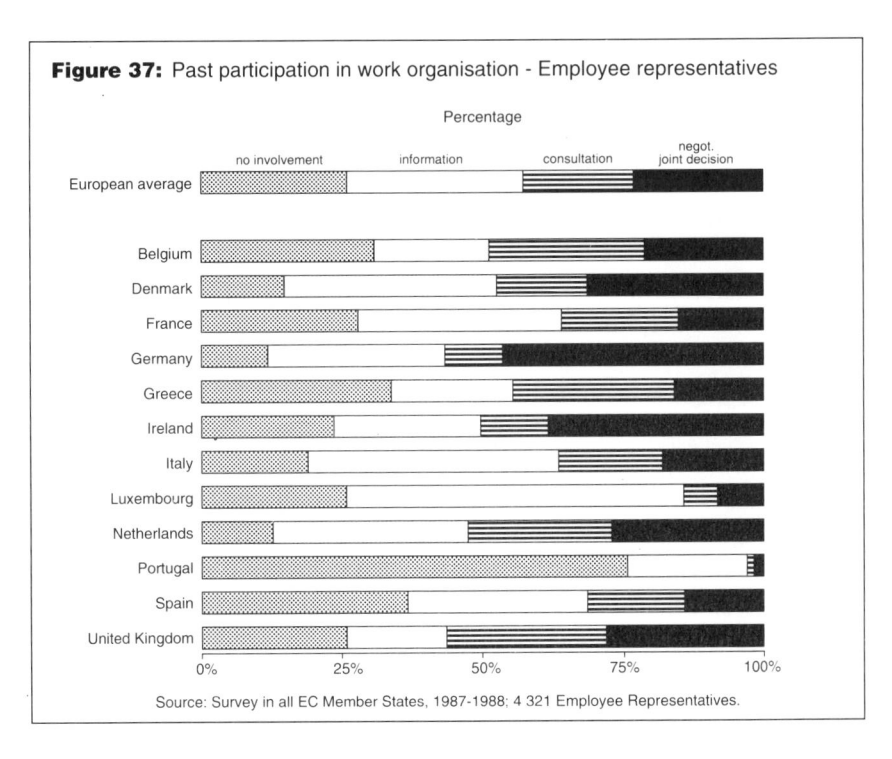

Source: Survey in all EC Member States, 1987-1988; 4 321 Employee Representatives.

In **Germany** both managers and employee representatives report the fewest cases of no involvement (9% and 12%) and the highest levels of negotiation/joint decision: around one out of two respondents in this country state that some form of co-determination characterises participation in work organisation. This can be partly explained by the characteristics of its industrial relations system and we know that unions have taken a strong interest in job design there.

As expected, **Denmark** fares very well: it has one of the lowest levels of no involvement (14% according to managers) and the third highest level of negotiation and joint decision: more than a third of managers report that type of involvement. The opinions of both sides of industry appear particularly close on these points. A good performance was only to be expected for the same reasons as in **Germany**.

In **Ireland**, according to three out of four managers, participation in work organisation takes the forms of consultation or higher levels — the third highest percentage in our sample. **Ireland** also occupies the second highest place according to employees' representatives (31%), the third according to managers, as regards the importance given to forms of co-determination in work organisation. Given the lack of adequate explanation for the high ranking of **Ireland** here, it is important to emphasise that there is some doubt about the method of sampling for the **Irish** results.

Belgium is characterised by the third highest level (64%) of forms of participation above information if we look at the answers of managers. It is very close to **Denmark** in this respect and this is also the case if we turn to the answers of employee representatives who report less (48%) consultation and co-determination than managers. The importance of negotiation and joint decision is comparatively high (30%) according to managers but slightly below average (21%) according to employee representatives.

The **Belgian** results might be explained by noting that in that country regulations relating to the powers of works councils require managers to provide information and to consult the works council whenever there are changes in work organisation. Moreover, when new technologies are introduced, the national collective agreement no.39 specifically requires information and consultation concerning changes in work organisation before the implementation stage. It is well known that this agreement is not always adhered to and is partially redundant with other regulations. But, given their existence, one could expect managers to have here a more positive view of their application.

The **Netherlands** comes next as far as the importance given in our survey by managers to negotiation and joint decision is concerned and only one out of ten managers reports the absence of any involvement of employees in these matters. There is a high degree of agreement between both sides of industry. This could also be expected: unions in the **Netherlands** have taken a strong interest in job design, although to a lesser extent than in **Denmark** and **Germany**. We know that in the **Netherlands** all our five explanatory factors are favourable and that a lower ranking than **Germany**, for instance, might be explained by the legislative provisions which give fewer co-determination rights.

In the **United Kingdom**, for one interviewee out of four, negotiation or joint decision takes place about work organisation. This is close to the European average. However, managers and employee representatives differ as to the extent of no involvement in the field of work organisation: over 26% of employee representatives report this situation while the corresponding figure for managers is half that. Could it be that the existing evidence of a strong dependence on the part of management on the co-operation of its workforce in the implementation stage of technological introduction distorts their view of reality? Another possible explanation for that pattern is a different appraisal by both sides of industry of what a minimum level of involvement is. As we know, prior to the beginnings of the 1980s shop stewards in Britain had traditionally built up a great deal of influence over various aspects of their work environment. Much of this influence was based on informal understandings between workers and management, established by long-standing customs. This could lead to different views of participation by management and by employee representatives.

Among the countries where the importance of the two lowest levels of involvement is still above average, **France, Spain** and **Greece** are comparable. According to managers, the extent of no involvement in these three countries varied between 29 and 36% and the employee representatives had practically the same views. In these three countries, forms of co-determination are reported by a proportion of interviewees varying between 12% and 16%, figures very close indeed. Given what we know about their industrial relations systems, these results below average are not really surprising, except possibly for **France**. There is considerable scope for participation in **France.** There, employers are required to inform the works councils of impending major changes in the enterprise which could affect working conditions. Moreover, one could expect the *"groupes d'expression directe"* brought into existence by the Auroux reforms and explicitly authorised to discuss only questions of work organisation, to have given some impetus to the discussion of these questions by worker representatives. We are certainly confronted here with the limits of the efficacy of legislative reforms.

We would expect **Italy** to be in the same group of countries and indeed, its figures are largely comparable as far as the importance of levels of participation above information are concerned. Yet, by comparison with these southern countries, **Italy** is conspicuous in having a small percentage of interviewees, close to the situation prevailing in **Denmark**, who report no involvement of employee representatives in work organisation and, at the same time, the highest level of information only according to managers: one out of two reports that situation. 45% of employee representatives give the same answer. In contrast to the French situation, this might be a case where practice is in line with the legislation : the 1970 Workers' Statute provides for a legal right to insert clauses into collective agreements which oblige **Italian** employers to inform the employee representatives of their investment plans and long term business policies which affect the terms and conditions of employees.

In **Luxembourg**, we also find information as the most common type of participation: 43% according to managers and 60% according to employee representatives but it is its low level of negotiation and joint decision which is remarkable: below 8 % according to managers and only 3% for employee representatives. Yet we notice that the amount of consultation reported by managers (21%) is close to the European average of 23%, while the employee representatives reported only 6%. Would this be a case of disagreement in the evaluation of the application of the right given to the *"comité mixte d'entreprise"* to be informed and consulted about new working methods?

Portugal must be set aside in the European Community : according to both managers and employee representatives, the absence of any involvement of employee representatives is, by far, the most common practice, amounting to three quarter of the cases, according to employee representatives. Correspondingly, it is in this country that the lowest levels of co-determination are reported, while consultation is a mode of involvement nearly non-existent. We already knew that participation in the introduction of new technologies is exceptionally low in **Portugal** by European standards but it is particularly striking in a matter which is of such direct concern to workers.

What are the expectations of our respondents about participation in work organisation in the future? At the European level, we mostly find a similar pattern for the two types of respondents. They both reject a situation where there would be no involvement at all, they want more consultation and, especially on the employees' side, less information. Yet, there is one striking difference: if managers also want more consultation and joint decisions on these matters, the proportion of employee representatives who share this opinion more than doubles to reach 60% i.e. a figure not much below the proportion of managers who want either consultation or some type of co-determination. It implies that the former also reject a situation in which only information would be given: 85% of them want more involvement than that. Obviously, participation in work organisation is important to employee representatives.

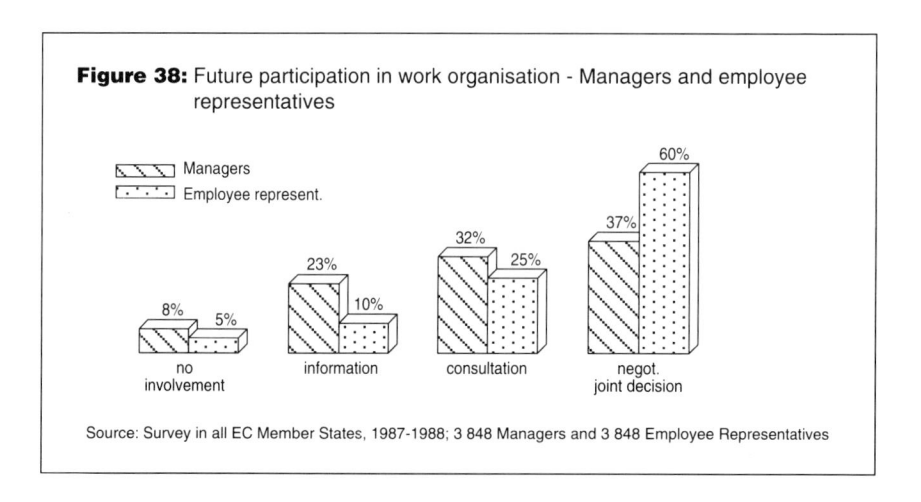

Figure 38: Future participation in work organisation - Managers and employee representatives

Managers: no involvement 8%, information 23%, consultation 32%, negot. joint decision 37%
Employee represent.: no involvement 5%, information 10%, consultation 25%, negot. joint decision 60%

Source: Survey in all EC Member States, 1987-1988; 3 848 Managers and 3 848 Employee Representatives

This pattern is present in every country, both for managers and for employee representatives. One exception is to be found in **Ireland** where managers want less consultation. However, there are differences in the amount of change wanted by respondents both between countries and between managers and employee representatives. We will comment here only on the points which modify or refine the analysis conducted about past participation in work organisation.

If we rank countries according to the degree of preference given by managers to forms of co-determination, the ranking of countries in the future is similar to past participation. There are, perhaps, two significant differences, one of which concerns **Germany**, and the other **Italy**.

Germany falls from the first position, now occupied by **Denmark**, to the third. The score is still quite high as 56% of German managers foresee more co-determination in the future. But, given that managers report in 45% of the cases that co-determination already exists in work organisation, there is little change by comparative standards. It should also be noticed that employee representatives' expectations rise to the first place among the countries surveyed: nearly eight out of ten prefer

PARTICIPATION IN WORK ORGANISATION

Figure 39: Future participation in work organisation - Managers

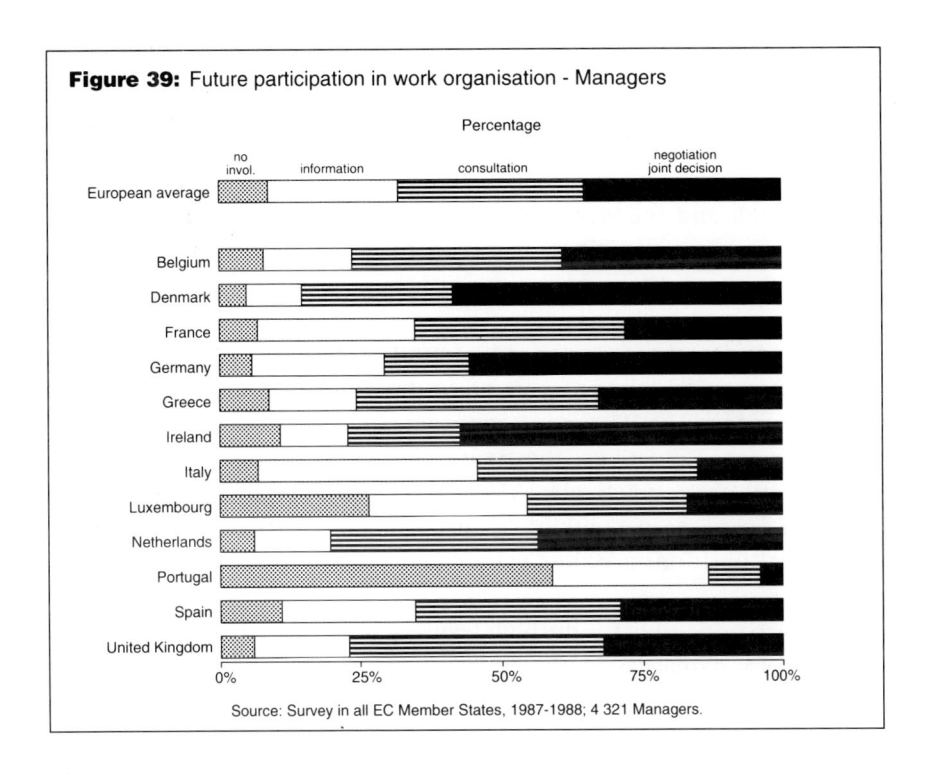

Source: Survey in all EC Member States, 1987-1988; 4 321 Managers.

Figure 40: Future participation in work organisation - Employee representatives

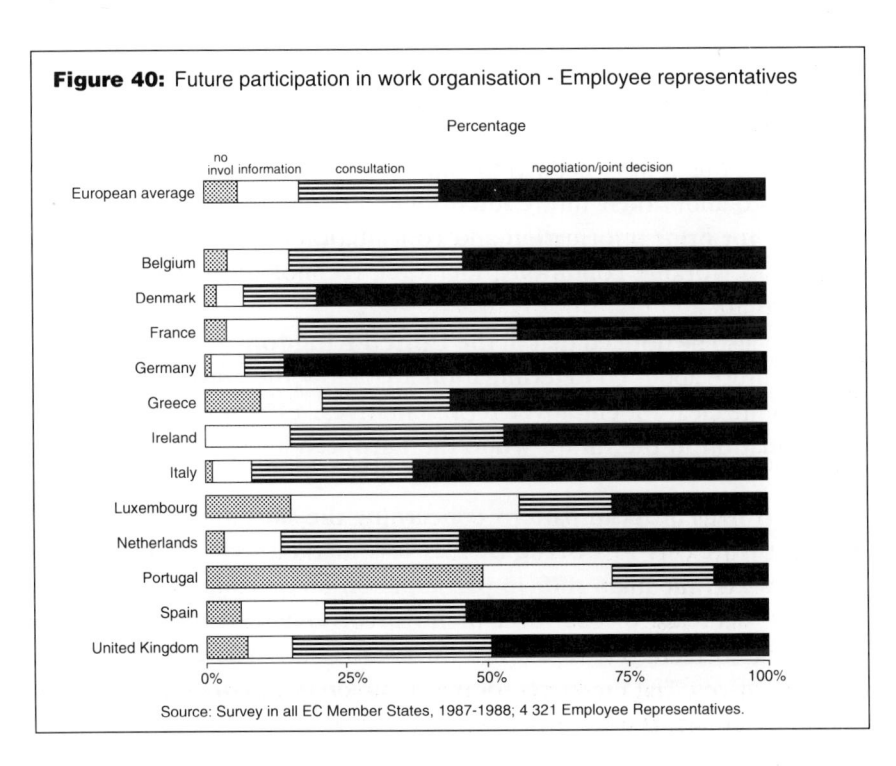

Source: Survey in all EC Member States, 1987-1988; 4 321 Employee Representatives.

negotiation or joint decision in the future. So, in an area where the levels of participation are the highest by European standards, the discrepancies existing between the expectations of both sides of industry leave room for conflict.

In **Denmark** and the **Netherlands**, especially in the latter, the other countries where participation in work organisation appears most developed, these discrepancies are not that important.

A very wide divergence is also to be found in **Italy**: the managers hardly change their minds about forms of co-determination in the future and are followed only by Portuguese managers in this respect contrary to the employee representatives: the figures for these two groups of Italian respondents are respectively 15% and 63% which means the eleventh and third positions in a ranking of European countries on this point. The conflictual positions are thus even clearer than in **Germany**.

Italy is also noticeable for the high levels of information and consultation expected by managers in the future. This remains in line with their assessment of the existing situation at the time of the survey.

In **Greece** there is a sizeable shift between past and future for both managers (from 15% to 32%) and employee representatives, but much more marked for the latter (from 16% to 57%). The two groups move up in the ranking of European countries, in this respect. It is possible that the 1988 works council law, not yet effective when the survey was conducted, might have been in the minds of the respondents when they were asked about their future intentions. Among other points, this law provides for prior information and consultation on improvements and changes in working conditions, including technological change.

It should also be noticed that in the **United Kingdom**, the expectations for the future about co-determination are below the European average. This is quite clear for employee representatives: 49% of them expect co-determination in the future while the European average is 58%.

Finally, if the European pattern concerning the shift between past and present also exists in **Portugal**, it is less marked than in most other countries and the main shift in participation levels from past to future is in the decrease of the level of no involvement: from 68 to 59% according to managers and from 76 to 48% according to employee representatives. The prospects for participation in **Portugal** are not very bright, even in this area of workers' concern.

8.4 Conclusions

The main conclusion to be drawn from this country analysis of the involvement of employee representatives in work organisation is that it reflects the varying degrees of importance attached to it in particular Member States. In the northern European countries such as **Denmark, Germany,** the **Netherlands** and **Belgium** work organisation assumes much more importance than is the case in the Mediterranean countries, where little priority is given to work organisation.

There is a cluster of countries for which no involvement is low and co-determination is high: **Germany, Denmark** and **Ireland**. The absence of involvement is also low but the amount of co-determination less prominent in the **Netherlands, Belgium** and the **United Kingdom. Luxembourg, France, Spain, Greece** and most noticeably, **Portugal** have much lower levels of participation in work organisation.

The expressed preferences of both sides of industry on this issue give some hope for enhanced participation levels in these matters in the future. Some degree of involvement is expected in all Member States, except in **Portugal**, and to a much lesser extent, in **Luxembourg**. It seems that negotiation and joint decision will substantially increase, but there is here a clear discrepancy between the expectations of managers and those of employee representatives. This discrepancy appears particularly high in **Germany** and **Italy** and might augur for some conflict between both sides of industry.

The importance of the results above cannot be overemphasised, particularly the low levels of participation in some countries. What is the purpose of participation as far as employee representatives are concerned? Participation is largely meaningless unless it is about work organisation; it is central to all of the trade union concerns.

CHAPTER NINE

Participation in Investment in New Information Technology

9.1 Introduction

During the late 1970s there was growing concern within the trade union movement in many European countries with the issues that were raised by new computing and information technologies. Unions emphasised in their policy statements that if the adverse implications of the new information technology were to be avoided, then unions would have to seek more effective ways of negotiating its introduction. As one British trade unionist put it in 1980:[1]

> "The trade union response to technological development is clearly linked to the area of industrial democracy. We are seeking to both widen the agenda on which bargaining takes place and to exert influence at a much earlier stage in decision-making".

This policy of emphasising the need to move beyond bargaining over the *effects* of technical change in the workplace (job security, payment levels, reduction of working hours, skills and health and safety etc.) and the need to consider in addition the *control* and *strategic issues* that were involved, found its way into some of the recommendations made by unions for negotiators of 'New Technology Agreements'. Trade unions emphasised that procedures for change should be based on collective bargaining and joint decision-making whose progress would be monitored by joint bodies; consultation with trade unions should begin prior to the decision to introduce new information technology; information and management plans should be made available to the unions and until such time as agreement was reached, the status quo should be observed.

By the mid 1980s the new technology agreement initiative had lost its impetus, and all the evidence suggests that of the few new technology agreements that were negotiated, nearly all fell short of their original trade union objectives of extending the ambits of collective bargaining to incorporate the strategic aspects of technological change.

Nevertheless, many trade union organisations in Europe still advocate the need for participation in the planning phase of new information technology introduction. The ETUC, for example, at its Sixth Ordinary Congress in Stockholm in 1988 called for:

(a) "the right of employee representatives to be fully informed, consulted and also to negotiate on all important company matters before decisions are taken;

(b) equal participation by employee representatives in all company decisions of significance to the workforce;

(c) extension of decision making rights at all levels of decision making according to the organisation of companies ... The employee representatives at all plants must accordingly ... have the right to be informed and consulted on company planning, to negotiate and to represent their interest jointly at the European level."

The rationale for trade unions to be directly involved in strategic issues concerned with investment in new technology is that the later bargaining is left in the process of change, the more difficult it becomes to achieve trade union objectives because management will already have taken major strategic decisions such as which type of equipment or system to install. By exercising influence by becoming a party to the decision-making process, unions would thus be able to play a proactive role and directly influence the impact of change on the workforce.

Such approaches however, are not without their problems. In some trade union circles — particularly so in some countries such as **France** and **Italy** — such involvement is regarded with suspicion. Many active trade unionists still believe that trade union representatives on joint committees or even company boards would not only have a negligible influence, but also their presence on such bodies would allow management to claim union support for its decisions. In the case of technological change this would require: the development of independent views on new equipment and systems to enable representatives to present alternatives to management proposals where appropriate; procedures to mobilise union support and maintain the accountability of representatives on participatory bodies; and organisational and other resources to facilitate these developments.

Even if these conditions were met, any influence trade unions had would have to be linked to support from the trade union outside the workplace in terms of full-time officer support and the education of workplace representatives in new technology and the options open to them. This in turn requires a prior union investment in the training of its full-time officers, and perhaps the development of specialist skills and expertise in aspects of technological change. Thus we would expect that countries where the trade unions were well-resourced and where research centres had been established to provide specialist expertise and support in new technology (**Denmark, Germany** and the **Netherlands**) would have the greatest degree of participation in investment criteria concerned with new technology. However, because participation in investment strategy is concerned primarily with the **planning** phase of technological change, we would not expect very high levels of participation since management sees this aspect of technological change as essentially their prerogative.

9.2 Participation in investment at the European level

Figure 41: Past participation in investment strategy - Managers and employee representatives

- no involvement: 59%
- information: 28%
- consultation: 6%
- negot. joint decis.: 7%

Source: Survey in all EC Member States, 1987-1988; 3 848 Managers and 3 848 Employee Representatives.

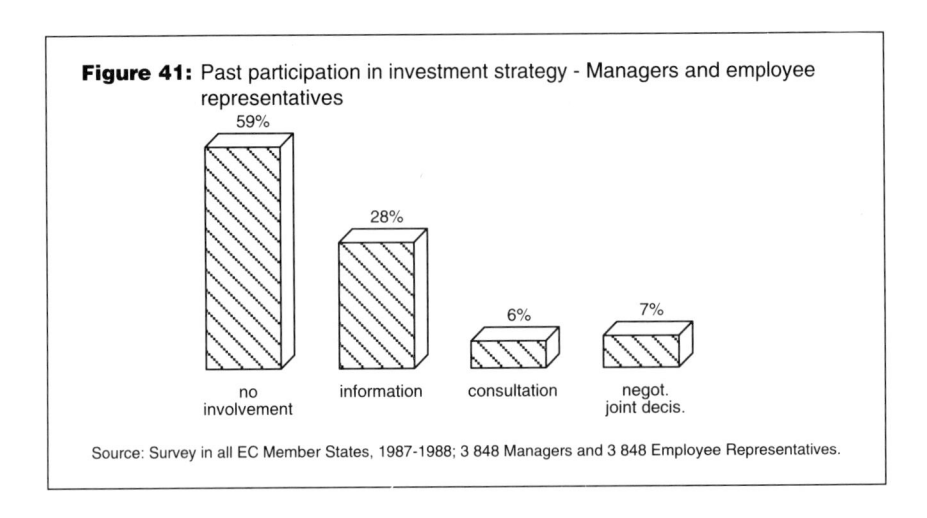

Figure 41 shows the combined responses of managers and employee representatives on the levels of participation in investment criteria for new information technology at the European level. As expected, employee representatives' involvement in investment strategy is low since participation here takes place in the **planning** phase. Well over half of the respondents state that there is no involvement at all in this area and that involvement is mainly limited to providing information. One out of four respondents reports that participation is characterised simply by the provision of information and just over one out of ten respondents report higher forms of participation. Consultation and co-determination/joint decision-making are given roughly the same importance in European firms.

9.3 Participation at Member State level

How do the countries compare in the survey results? If we consider the responses of managers, we find that "no involvement" is at its lowest in **Denmark** and **Germany** where two thirds of respondents report that some form of participation takes place. These two countries also have the highest levels of negotiation/joint decision-making among the twelve European countries, respectively 15% and 16%.

The **Netherlands** comes next with 55% of managers stating that there is at least information given to employee representatives on investment strategy in their firms. This percentage falls below 50% for the nine remaining countries in the Community.

Figure 42: Past participation in investment strategy - Managers

Percentage

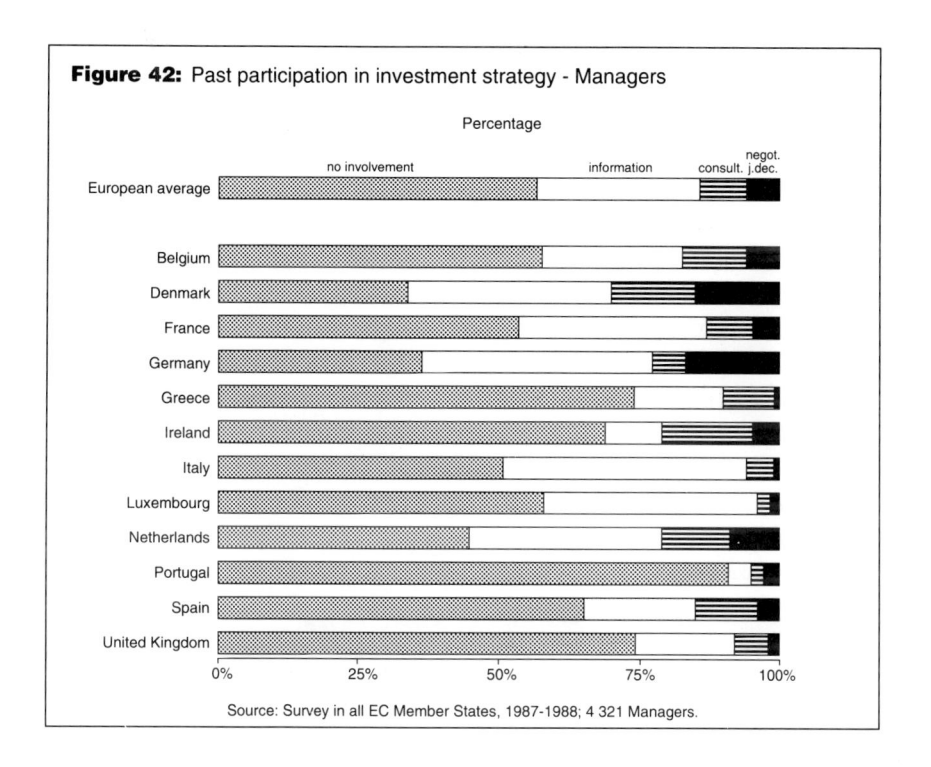

no involvement | information | consult. | negot. j.dec.

- European average
- Belgium
- Denmark
- France
- Germany
- Greece
- Ireland
- Italy
- Luxembourg
- Netherlands
- Portugal
- Spain
- United Kingdom

0%   25%   50%   75%   100%

Source: Survey in all EC Member States, 1987-1988; 4 321 Managers.

Among those, **Italy** is best placed: 51% of respondents there report the absence of any participatory involvement and 42% answer that information is provided, a level just above the one obtained in **Germany** (41%) and **Denmark** (36%).

The countries where participation in the investment criteria of new technology falls to its lowest levels — less than one respondent out of four — are **Greece**, the **United Kingdom** and **Portugal**. For the two Mediterranean countries, this result falls broadly in line with the pattern of participation we have seen there so far. This is not the case, however, in the **United Kingdom**.

Figure 43 illustrates the responses of employee representatives on the levels of participation in existence at the time the interviews were carried out. There is a striking similarity between the pattern of responses of both sides, although employee representatives tend to be marginally more sceptical about the levels of participation in existence than managers. The same three countries — **Germany**, **Denmark,** and the **Netherlands** — again appear to have the highest degree of negotiation/joint decision-making as well as the lowest degrees of "no involvement". Employee representatives again give similar rankings for the other countries, and thus we can be confident about the validity of our data.

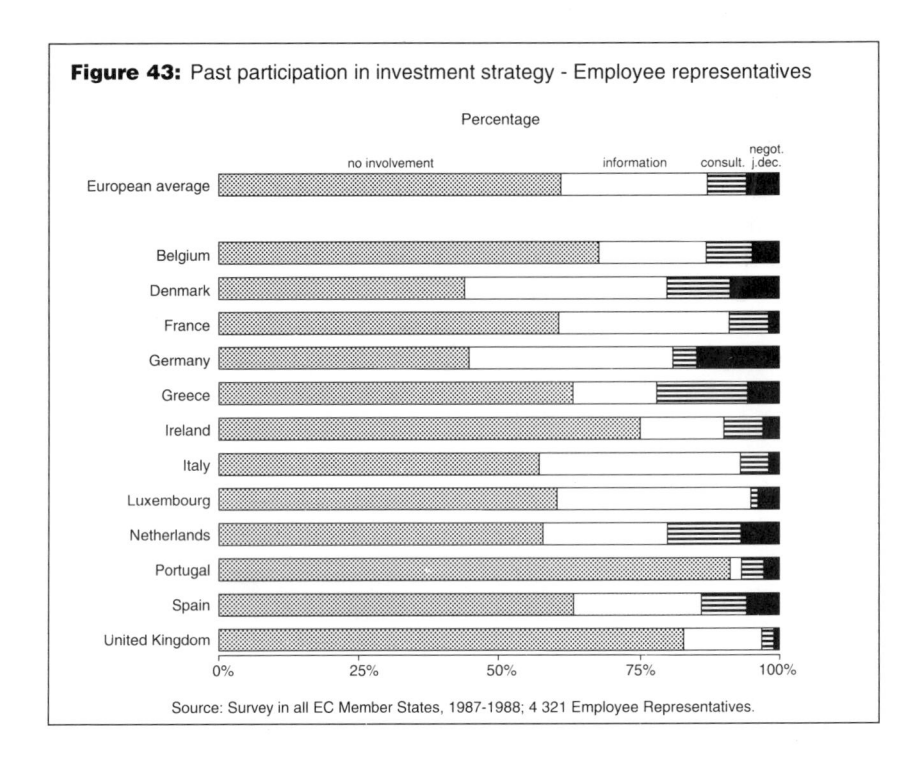

Figure 43: Past participation in investment strategy - Employee representatives

[Bar chart showing percentages of no involvement, information, consultation, and negotiation/joint decision for European average, Belgium, Denmark, France, Germany, Greece, Ireland, Italy, Luxembourg, Netherlands, Portugal, Spain, and United Kingdom.]

Source: Survey in all EC Member States, 1987-1988; 4 321 Employee Representatives.

How can these rankings be explained? If we once again return to our explanatory factors we can offer some explanation for the relatively high German ranking by pointing out the legal regulations which give wide powers to Works Councils in **Germany**. In addition, managers there depend on the skills and problem-solving abilities of their workforces, and the choice of technological systems may be a matter that managers in those enterprises which have a more "open" management style are prepared to arrive at decisions jointly with employee representatives.

In **Denmark,** with its greater reliance on collective bargaining rather than on legislative provisions which empower Co-operation Committees, managers are prepared to negotiate and consult on a wide range of enterprise issues which affect employees. In the **Netherlands**, works councils have the right to be consulted on major investment decisions.

The results for the other countries fall generally in line with our expectations with the exception of the **United Kingdom**. Here, there is a particularly high level of "no involvement" and the amount of negotiation/joint decision-making is minuscule compared to other European Community countries (including **Portugal**). How can we account for this high level of "no involvement" in the **United Kingdom**? If we consider the five main factors which shape the opportunities for employee involvement in new technology introduction, it can be argued that four of them are unfavourable, and only the dependence of management on workforce co-operation in the implementation stage of technological introduction can be said to be favourable.

There is certainly evidence of a strong dependence on the part of management on the co-operation of its workforce in the **implementation** stage of technological introduction.[2] However, this means only a very loose form of **informal** participation by the workforce and it is often carried out as much to circumvent the influence of trade union representatives as it is to further genuine participation by employees.

At present, any idea of worker directors does not feature on the political agenda in Britain. British managers are firmly against any participation at Board level and are also opposed to greater participation by shop stewards at a lower level. There is evidence to suggest that **shop stewards' committees** sometimes function in similar ways to the **German works council**[3] but that, because management neglects to initiate early dissemination of information and consultation, they are unable to elicit the same involvement in the problems of change as was achieved in **Germany**. Even if management moved towards involvement in decision-making on a wider range of issues, local shop stewards' committees would not have the resources and capacities to respond effectively to such a challenge, nor would they achieve much support in this respect from the external trade union movement. In any case the bargaining power of trade unions in Britain has been considerably eroded during the 1980s by a combination of anti-union legislation, major changes in the labour market, the contraction of the manufacturing sector, the erosion of full-time male, manual employment, the growth of the service sector and the creation of a large number of part-time jobs.

9.4 Future participation in investment in new information technology

What are the prospects for the future? Figure 44 on p. 148 sets out the preferences of managers for the future at the European level insofar as investment criteria in new information technology are concerned, in the light of their past experience with participation. Despite the controversial nature of this issue, there is nevertheless a tendency for managers to prefer higher levels of participation in the future. The level of "no involvement" shrinks substantially from 56% to 32%; there is a marginal increase in the provision of information; one in five managers opt for consultation and those who prefer negotiation/joint decision-making doubles from 7% to 14%.

Figure 44: Past and future participation in investment strategy - Managers

no involvement: 56% past, 32% future
information: 30% past, 34% future
consultation: 7% past, 20% future
negot./joint decision: 7% past, 14% future

Source: Survey in all EC Member States, 1987-1988; 3 848 Managers

Figure 45: Past and future participation in investment strategy - Employee representatives

no involvement: 62% past, 17% future
information: 26% past, 23% future
consultation: 6% past, 25% future
negot./joint decision: 6% past, 35% future

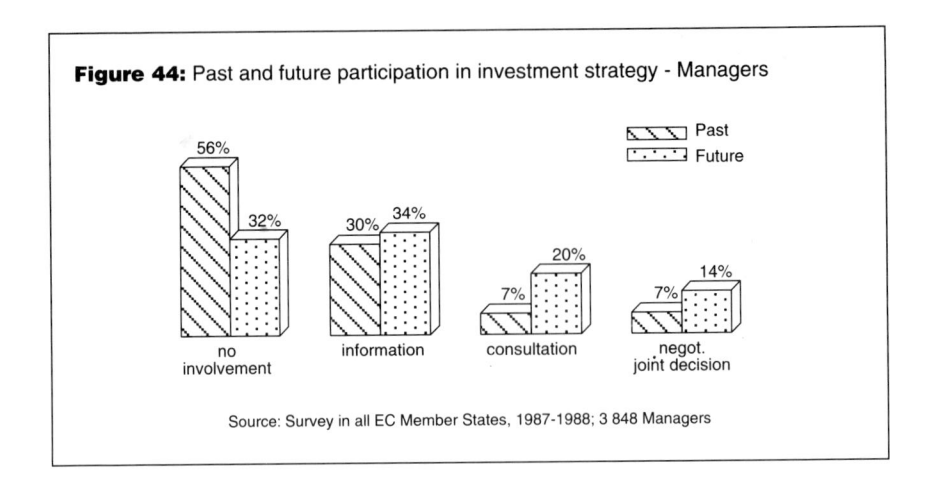

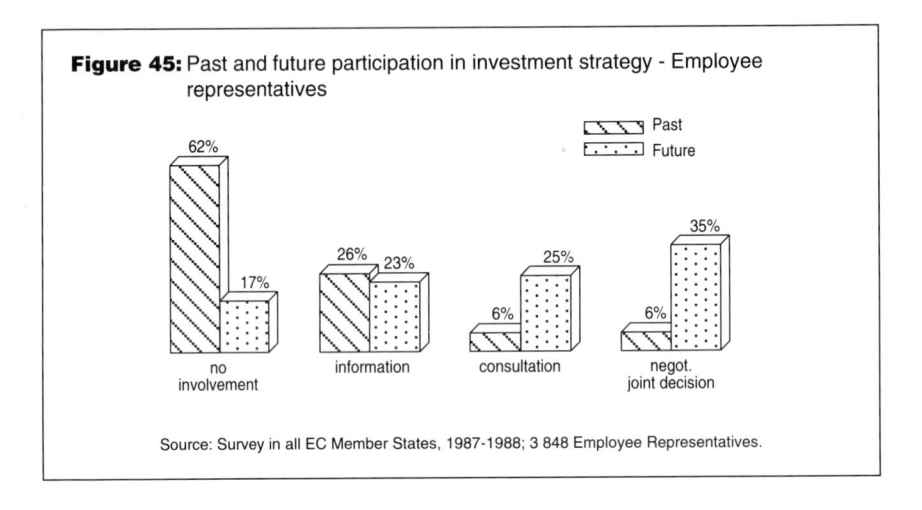

Source: Survey in all EC Member States, 1987-1988; 3 848 Employee Representatives.

Figure 45 above gives the preferences of employee representatives at European level for participation on this issue in the future. Not surprisingly, employee representatives expect higher levels of participation on this issue than their management counterparts. The preferred method is negotiation/joint decision-making (35%); one in four opt for consultation; the percentage of those preferring information slightly declines. Significantly however, the level of "no involvement" declines substantially.

What kind of variations do we see in individual countries in terms of future participation? Figure 46 on p. 149 portrays the kind of involvement that managers in each European Community country prefer employee representatives to have in new information technology investment for

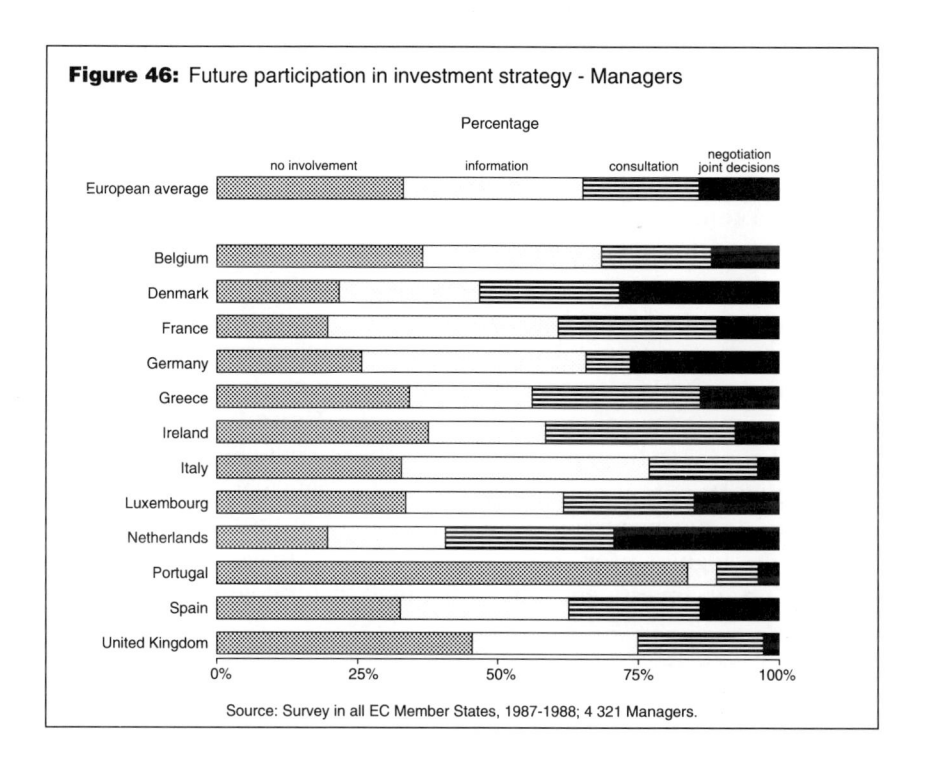

Figure 46: Future participation in investment strategy - Managers

Source: Survey in all EC Member States, 1987-1988; 4 321 Managers.

the future, based on their past experience. The responses of managers are particularly significant as a measure of the likely forms of participation in the future. After all, it is managers rather than employee representatives who are likely to to determine the agenda for participatory practices.

The first point to note is that in **every** country managers envisage higher levels of participation than at the time the interviews were conducted. Thus this area of technological change, despite its controversial nature, is also a matter where there is a distinct trend towards higher levels of participation across the Community.

The second point to note is that this trend towards higher levels of participation is particularly marked in the **Netherlands**, **Denmark** and **Germany**; it is in these three countries that managers expect the most co-determination: about one out of every four express that opinion although at the same time, comparatively very few German managers envisage consultation in investment criteria. Thirdly, in the other countries, co-determination is favoured at best by 15% of respondents and this proportion falls down below 10% in **Ireland**, **Italy**, **Portugal** and the **United Kingdom**.

Despite the general trend throughout the European Community for higher degrees of participation in the investment criteria for new technology, it is only in the **Netherlands** and **Denmark** that over half of the managers want consultation or higher forms of participation. There is a group of countries where those managers envisaging consultation or higher levels of participation cluster around the 40% level: in **Greece**, **Ireland**, **France**, **Luxembourg** and **Spain**. Somewhat surprisingly, in **Germany** and **Belgium**, this percentage is only around 30%. This unusually low position of **Germany** reflects the fact that consultation appears as the least desired form of participation while co-determination and, more so, information are preferred. Under half of all managers envisage at least consultation in **Italy**, **United Kingdom** and **Portugal**. The prospects for enhanced forms of participation in the **United Kingdom** and **Portugal** appear to be particularly bleak; the overwhelming majority of managers in **Portugal** and nearly half of the managers in the **United Kingdom** do not want any participation at all in the future. In **Italy**, we find the highest proportion — 44% — of managers ready to give information on investment criteria for new technology.

Figure 47: Future participation in investment strategy - Employee representatives

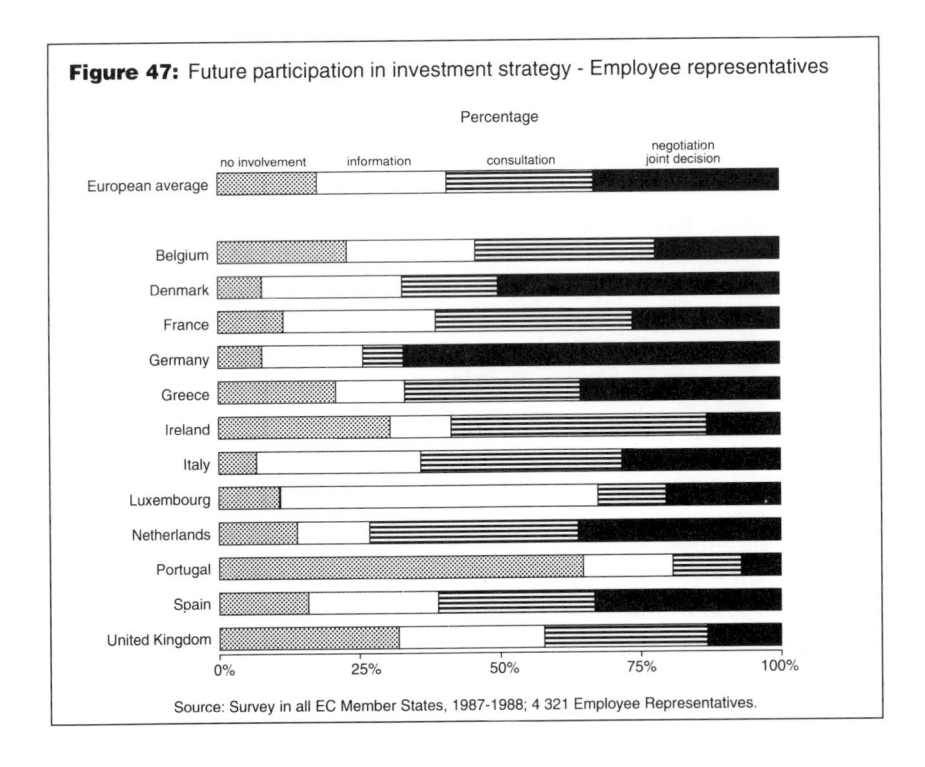

Source: Survey in all EC Member States, 1987-1988; 4 321 Employee Representatives.

Figure 47 above sets out the preferences of employee representatives for participation in each country in the future. As one would expect, employee representatives seek much higher levels of participation than their management counterparts in every Member State. More than half of them seek either consultation or co-determination in most countries, with the notable exceptions of the **United Kingdom**, **Luxembourg** and **Portugal**; this is consistent with managers' positions in these three countries.

It is particularly striking to note that nearly three out of four employee representatives in **Germany** expect negotiation or joint decision-making as the primary form of participation. The higher expectations of employee representatives here may reflect the importance that the DGB in **Germany** places on seeking a strong influence in all aspects of technological change; indeed the German unions have spearheaded the policies of the ETUC in this area. It is also worth noting that there is a wide divergence between the views of German employee representatives and their management counterparts; 67% of German employee representatives seek negotiation/joint decision-making in the future compared with 26% of German managers.

The gap between the expectations of the two sides of industry is comparatively narrow in **Denmark**, a country where consensual relations are dominant, and as wide in **Italy** as it is in **Germany**. Some problems might appear in the future there. The employee representatives' expectations in **Ireland** and **Belgium** are more in line with those of managers.

9.5 Assessing the results in the light of the Val Duchesse Joint Opinion

Finally, how can these country survey results be assessed in the light of the **Val Duchesse** Joint Opinion issued by the working party on 'Social Dialogue and the new technologies' concerning training and motivation, and information and consultation? If we take the benchmark that we used for assessing the results on participation in training (Chapter 6) employee representatives must be, at the minimum, consulted about investment criteria in new information technology. The mere provision of information on this topic does not, therefore, fall within the guidelines laid down in the joint opinion. As we have already mentioned, levels of participation involving negotiation and/or joint decision-making about training are superior to the Val Duchesse provisions.

If we take the responses of managers at the time the interviews were conducted (Figure 42), there is not one European Community Member State which falls in line with the Val Duchesse guidelines. In **Denmark** we find that just over one in four managers report that employee representatives are at the minimum consulted about investment criteria for new information technology, but most other countries are well below this level.

However, there are more favourable prospects for consultation in the future. Figure 46 on p. 149, which sets out managers' preferences for the future, shows some measure of improvement; in two countries, **Denmark** and the **Netherlands**, a majority of managers state that they intend to either consult with employee representatives or initiate higher levels of participation in new information technology in the future. There are also improvements in the other countries, but by no means to the same degree. Around 30% of managers in most Member States are likely to be following Val Duchesse guidelines in the future.

9.6 Conclusion

It is clear from our survey data on participation in investment criteria for new information technology that the levels of participation are very low indeed. There is nothing surprising about these results because this aspect of technological change in enterprises is concerned with the **planning** phase, and management has always seen decisions of this kind as their prerogative. As one would expect, where participation does occur, it takes place in those countries where the trade union movement places a priority on such issues — in **Denmark, Germany** and the **Netherlands**.

Despite the fact that the ETUC policy is to ensure that all aspects of technological change are the subject of negotiation or joint decision-making, their success in achieving this objective has so far been minimal. There is evidence that management is prepared to take a more accommodating view about involving employee representatives in investment decisions in the future, but only in two countries — **Denmark** and the **Netherlands** — do we find a majority of managers who envisage that the Val Duchesse guidelines of at least consultation are likely to be observed.

As we mentioned in the introduction to this chapter, the involvement of employee representatives in such strategic areas of management decision-making about technological change is not without its problems for the labour movement. This too is reflected in our results as we see those countries where the labour movement is ambivalent about such involvement or is split across religious or political lines e.g. in the **United Kingdom** (where some trade unions have feared that such involvement in management decision-making cuts across collective bargaining) and in **France** and **Italy** (where there are traces of an ideological hostility to taking part in crucial management decisions).

Our survey results also reflect the general trend throughout Europe for management to seek much more decentralisation and de-regulation in industrial relations. In many enterprises strategic decisions about investment in new technology are taken at higher levels in the company which are far removed from the establishment itself and this severely limits the scope of trade union influence at the level of the establishment. Our survey results must also be seen in the light of the general trend throughout Europe in the 1980s for increased flexibility.

This drive for more flexibility means that management are placing much more emphasis on the operational level. In other words, management is placing emphasis on employee involvement in the **implementation** phase of technical change. This drive on the part of management however, is a form of *management-initiated* participation where management deal with workers directly, ignoring or cutting across employee representatives by means of quality circles, teamworking, briefing groups, video presentations etc. whose purpose is to inculcate a greater sense of involvement on the part of individual workers into the aims of the enterprise. Within the European Community there is evidence that many employees welcome such initiatives and we can observe a growing demand for and experience with the involvement of workers in quality circles and the like (especially in **Belgium, France,** the **United Kingdom**); more rarely, forms of profit-sharing (**Denmark**); and the distribution of shares to employees (**Germany,** the **United Kingdom** and **France**).

Compared with other aspects of participation in new information technology so far such as training (chapter 6), health and safety (chapter 7), work organisation (chapter 8), our survey results show that levels of participation in investment decisions concerned with technological change are very low indeed.

---

1. David Lea, chairman of a special employment and technology committee established by the British TUC.
2. Christel Lane, *Management and Labour in Europe,* Edward Elgar, 1989, pp. 189-195.
3. Bessant, J.R. and Grunt, M., *Management and Manufacturing Innovation in the United Kingdom and West Germany,* Gower, 1985.

Chapter Ten

Participation in Product Quality

10.1 Introduction

One of the primary reasons for the creation of the Single Market in the European Community was to enable European enterprises to meet the challenge of competition in world markets from countries such as Japan and some nations in the Far East. Central to the achievement of this objective is the need to provide products and services on world markets which meet the needs of consumers; the quality of such products and services is therefore of prime importance for the success of European enterprises. Indeed, the provision of high standards of products and services is in the interests of both managers and employees.

The apparent success of companies in Japan and the Far East in avoiding industrial conflict and achieving high levels of worker commitment to quality and consistency of output has had a major impact on current managerial thinking in Europe. Management in many European enterprises have adopted several aspects of Japanese managerial practice and this has led to an emphasis on the links between greater status equality, labour flexibility and a sense of personal commitment to the quality of finished products and services. This change in managerial thinking is itself linked to two strands of managerial thinking which gained prominence during the 1980s; total quality management and human resource management.

Total Quality Management (TQM) involves placing an emphasis on quality in all aspects of the functions of the enterprise; in recruitment, training and development, managerial practice etc. as well as in the final product or service. It involves placing the needs of the customer first and all employees are expected to subscribe to this philosophy. The enterprise has a "mission statement" and management seeks to inculcate employees into the "culture" of the firm. During the 1980s quality circles were set up in many enterprises across Europe. Quality circles are voluntary groups of between six and eight people from the same workplace who meet for an hour each week under the leadership of their supervisor to solve work-related problems in their department. Members of the circle select the problems they wish to tackle; collect the necessary data; apply systematic problem-solving techniques in working towards solutions; present their findings and proposed solutions to management for approval; implement their solutions where practicable, and monitor their effect.

Such direct or task participation on the part of individual employees has met with a mixed response from trade unions. On the one hand they are welcomed because they can improve job satisfaction and provide more control by employees over their jobs; on the other hand they are opposed because they are seen as undermining trade union influence on the shopfloor by bypassing forms of collective representation. Occasionally, shop stewards or workplace representatives exercise a leadership function for the quality circles. However, the evidence of case studies and surveys suggests that quality circle programmes often fail not because of employee or trade union opposition, but because of a lack of *management* support, the poor response to circle initiatives from management, a closed management style, and lack of recognition given to circle activities.[1]

The growth of Human Resource Management (HRM) is linked with the philosophy of Total Quality Management in its approach to personnel policy. While there is some doubt about its precise meaning, a distinctive feature of this approach is the degree to which it focuses on the connections between overall company strategy and personnel policies. The emphasis in human resource management on linking employee relations with the establishment and maintenance of a company culture is relevant to work organisation, training, making the best use of new technology and the quality of products and services — all of which are subjects covered by our survey.

Both of these strands are linked to the popular success among many European managers of Peters' and Watermans' book *In Search of Excellence* in 1982.[2]

While these changes in managerial thinking during the 1980s have tended to emphasise the "commonality of interests" between managers and employees, all this is not to say that there are no differences of interest between the two sides. If this were so, the need for participation would not arise. What then, are the differences of interests between managers and employees' representatives?

For **managers** of a firm operating in a competitive environment, product strategies have to fit in with market constraints; investment in technology has to follow both product design (see Chapter 9) and competitive pressures for low production costs; and work organisation has to be fitted to the adopted technology so that the resulting products and services can optimally exploit opportunities in markets. In such a situation, the utilisation of labour is determined by the requirements of an optimal organisation of work, and it is the task of collective bargaining and joint decision-making to ensure that the labour market supplies exactly what the firm needs.

For **employee representatives**, by contrast, their interests lie in the opposite direction. By acquiring control over the supply of labour through collective organisation, they seek to define wage levels, skill structures, work organisation, training requirements and other employment conditions to which firms must then adjust their organisation of work.

In other words, technology would have to be designed so as to be compatible with the structure of the labour supply and with the preferred organisation of work. In sum, to the extent that the resulting technology is dedicated to a particular category of products or services, employee representatives' interests require that product strategies be matched to technical choices, rather than vice versa. Employee representatives, unlike managers, are formally involved in a company's decision-making only as far as it relates to the company's external and internal labour markets. As a result, important parameters affecting the choice and design of new technology are outside their institutionalised sphere of influence; in the absence of any ability to negotiate about product market factors, employee organisations are therefore limited to creating institutional rigidities at the lower end of the managerial decision-making sequence that constrain management's decisions on pay, work organisation and in part, technology.

From the above discussion, it should be evident that the issues of training, work organisation, investment in new technology (Chapters 6, 7, 8 and 9) are all legitimately matters in which employee representatives can have a participative function. Training, work organisation, and product and service quality tend to be associated with the **implementation** phase of technological change; whereas, investment is largely concerned with the **planning** phase. As such we would expect our survey results for product and service quality to reflect this.

10.2 Past participation in product & service quality at European level

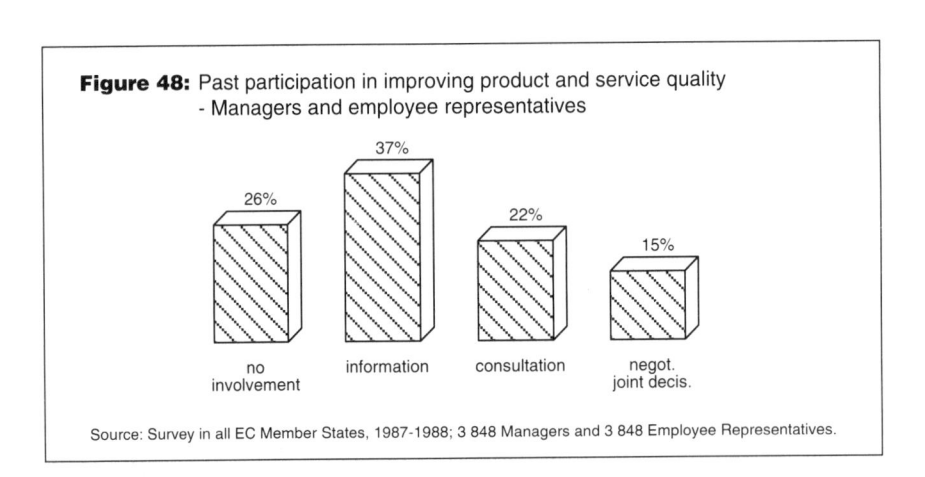

Figure 48: Past participation in improving product and service quality - Managers and employee representatives

no involvement	information	consultation	negot. joint decis.
26%	37%	22%	15%

Source: Survey in all EC Member States, 1987-1988; 3 848 Managers and 3 848 Employee Representatives.

PARTICIPATION IN PRODUCT QUALITY

Figure 48 on p. 158 shows the combined responses of managers and employee representatives on the levels of participation in product and service quality at the European level. The responses of managers and employee representatives differed hardly at all and therefore the results have been combined. Well over a third of respondents report that information is the form of participation which is most common, with just under a quarter of respondents reporting consultation. As we would expect from our discussion above, the level of negotiation/joint decision-making is quite low (15%); this represents twice as many as was the case for investment strategy. One in four report that there is no involvement whatsoever.

10.3 Comparative results in the 12 Member States

How do the countries compare with each other on this issue? Figure 49 above sets out the responses of managers on the levels of participation in existence in improving product and service quality at the time the interviews were conducted? The figure illustrates that the levels of "no involvement" rarely exceed 30% of respondents, and are particularly low in **Denmark, Ireland,** the **Netherlands** and the **United Kingdom.** Over three in ten managers in **Denmark** report that improving product and service quality is a matter for negotiation or co-decision, and coupled

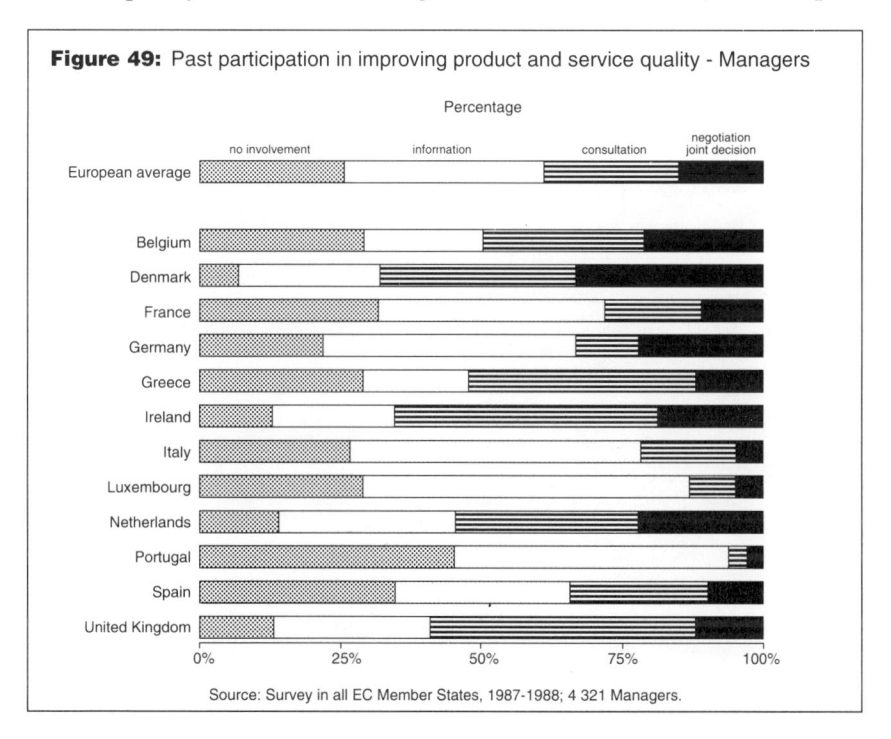

Figure 49: Past participation in improving product and service quality - Managers

Source: Survey in all EC Member States, 1987-1988; 4 321 Managers.

with consultation, it accounts for over 70% of cases. This indicates the wide range of issues which are dealt with by Co-operation Committees in that country. In the **Netherlands, Belgium** and **Germany** around 20% of managers state that product and service quality is a matter of negotiation or joint decision-making — perhaps reflecting the powers of works councils.

The results for the **United Kingdom** perhaps reflect the impact of the increased tendency of British managers to adopt HRM-type approaches to employee involvement during the 1980s — by using techniques such as teamworking, face-to-face communication programmes, briefing groups, quality circles and so on. Hence the comparatively high degree of consultation and information with respect to quality.

In line with our survey results in other aspects of participation, theresults for **Luxembourg** and **Portugal** are largely characterised by the mere provision of information.

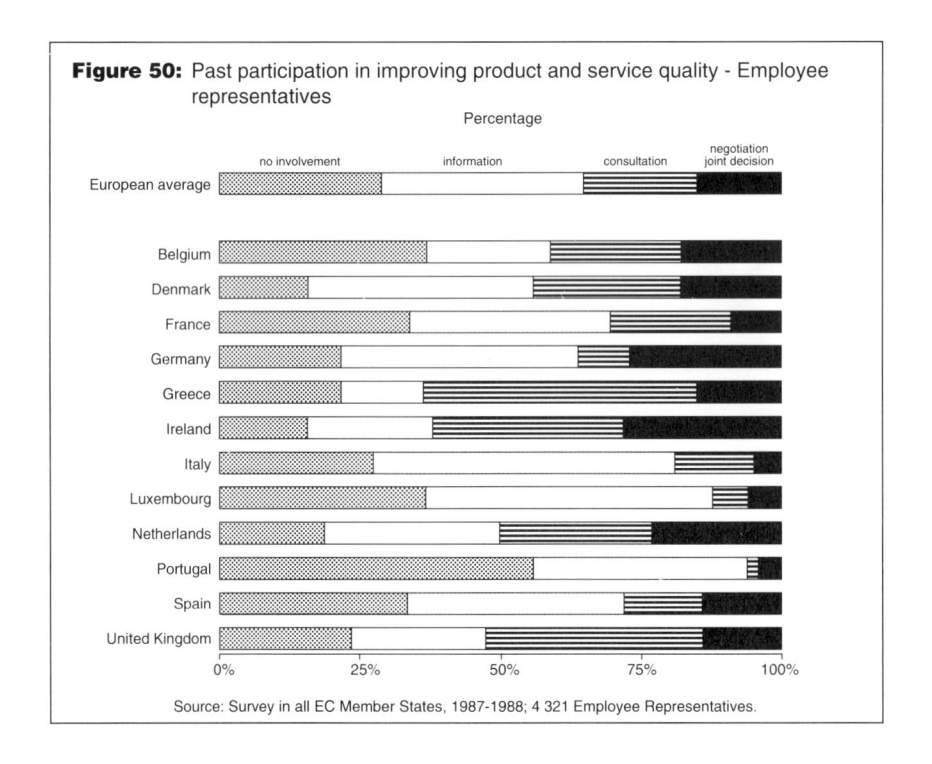

Figure 50: Past participation in improving product and service quality - Employee representatives

Source: Survey in all EC Member States, 1987-1988; 4 321 Employee Representatives.

The general pattern of the responses of employee representatives is rather similar to that of managers. As one would expect, employee representatives are marginally more sceptical about the levels of participation in existence at the time of the survey, although the differences overall are rather minimal. It is notable that in **Denmark** employee representatives are much more sceptical than their management counterparts; they claim that there is much less consultation and negotiation/joint decision-making and more cases of no involvement at all (44% of them report consultation and higher forms of participation while the comparable figure is 68% for managers). In the other European countries, the difference in percentage points between both sides of industry concerning the existence of consultation and higher forms of participation never exceeds 9%.

10.4 Future expectations of participation in product and service quality

Figure 51 above sets out the change that takes place at European level when the respondents were asked about their future expectations about participation in product and service quality. The responses of both sides have been combined together.

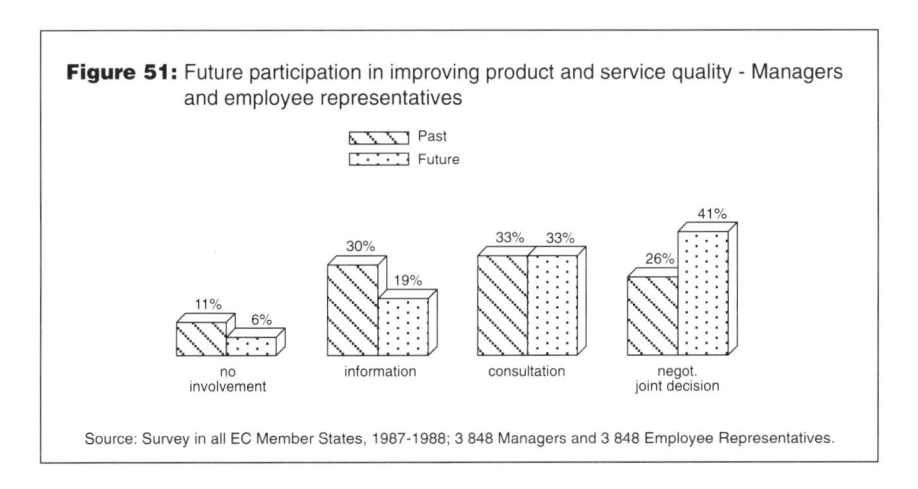

Figure 51: Future participation in improving product and service quality - Managers and employee representatives

	Past	Future
no involvement	11%	6%
information	30%	19%
consultation	33%	33%
negot. joint decision	26%	41%

Source: Survey in all EC Member States, 1987-1988; 3 848 Managers and 3 848 Employee Representatives.

There is a clear shift in attitudes towards a preference for much higher degrees of participation in this aspect of technological change. The percentage of those who reported that participation was characterised by "negotiation" or "joint decision-making" at the time the interviews were conducted (26%) shows a significant increase when respondents were asked about their preferences for the future (41%). The percentage of those reporting "no involvement" in the past shrinks when we consider future expectations. The percentage reporting consultation remains unchanged at 33%, and there is a decline in the percentage who reported that participation in technological change was characterised by the mere provision of information.

Figure 52: Future participation in improving product and service quality - Managers

Source: Survey in all EC Member States, 1987-1988; 4 321 Managers.

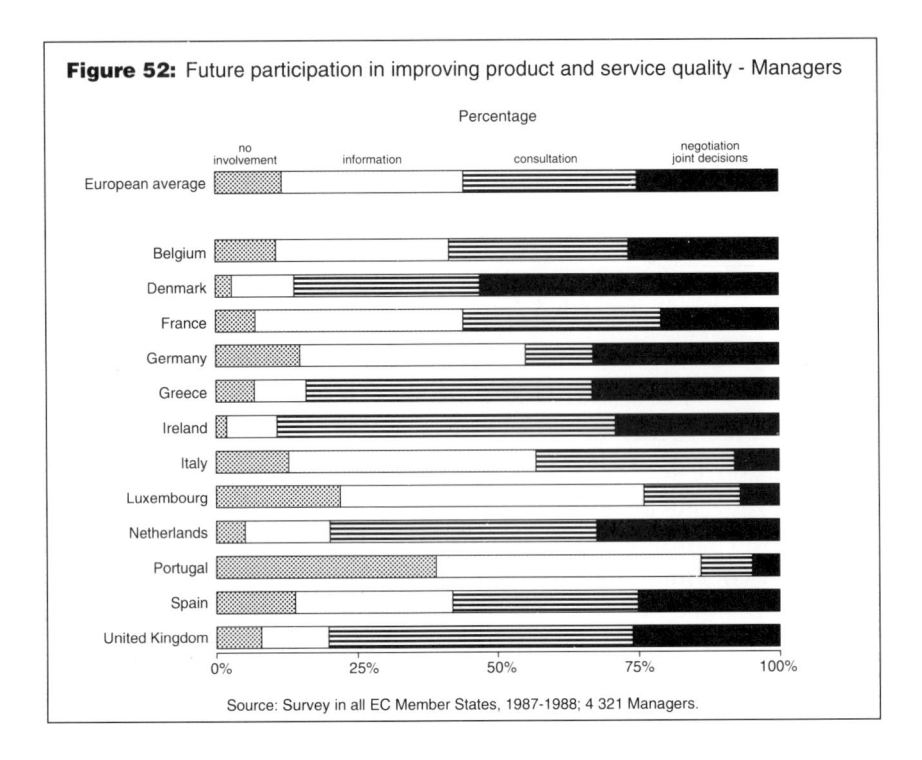

What are the prospects for the future regarding this aspect of participation? Figure 52 shows that, in keeping with other aspects of participation, there is a pronounced shift on the part of managers in all countries, albeit in different degrees, towards higher levels of participation in the future. Across Europe, the numbers of those managers who would prefer no involvement by employee representatives has shrunk substantially — in some cases (**Denmark, Ireland** and the **Netherlands**) to minuscule percentages. Moreover, around a quarter of managers in nine out of the 12 Member States (**Portugal, Italy** and **Luxembourg** being the exceptions), expect this aspect of participation to be the subject of negotiation or joint decision-making. Well over half the managers in the **United Kingdom** favour making participation in improving product and service quality the subject of consultation, again reflecting the importance of the changes in British managerial philosophy which has been particularly pronounced in that country. **Ireland** has even better results; the percentage of Irish management respondents who favour consultation or higher levels of participation almost reaches 90%.

In keeping with the findings of the survey generally, employee representatives are much more ambitious in their aspirations for the future than their management counterparts. Figure 53 on p. 163 illustrates the expectations of employee representatives in product and service quality

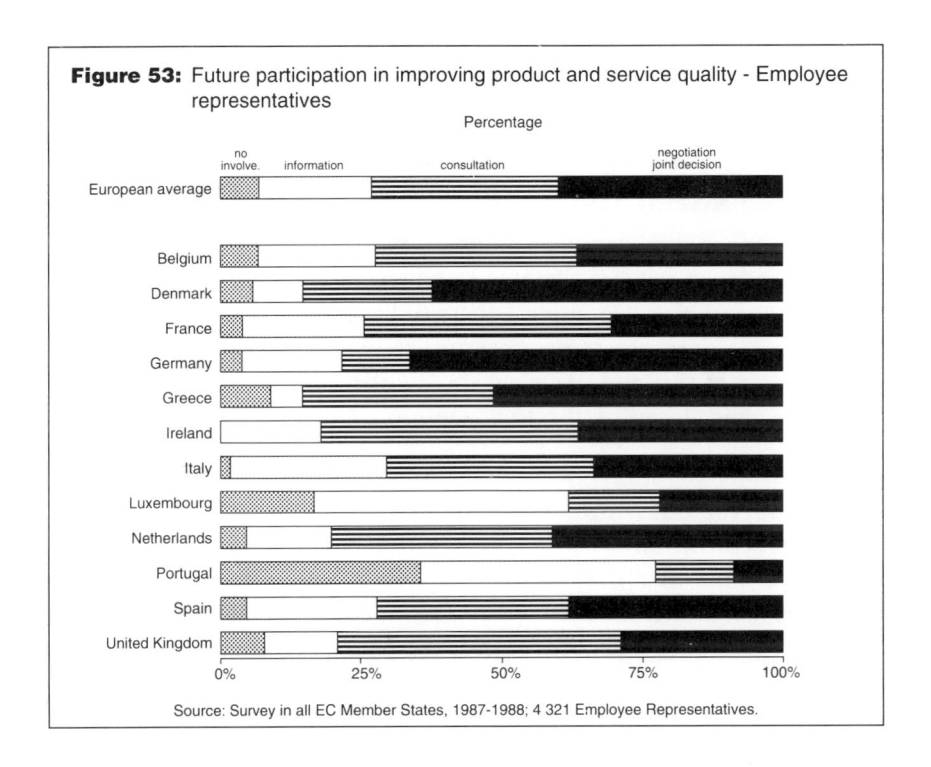

Figure 53: Future participation in improving product and service quality - Employee representatives

Percentage

Legend: no involve. | information | consultation | negotiation joint decision

European average	
Belgium	
Denmark	
France	
Germany	
Greece	
Ireland	
Italy	
Luxembourg	
Netherlands	
Portugal	
Spain	
United Kingdom	

Source: Survey in all EC Member States, 1987-1988; 4 321 Employee Representatives.

in the future. Again, it shows that in all 12 Member States employee representatives are very enthusiastic for higher participation levels.

Apart from **Portugal**, those who are prepared to tolerate no involvement at all is less than 10%. Around three in every ten respondents favour negotiation or joint decision-making. It is worth noting that employee representatives in **Denmark** and **Germany** are particularly keen to see product and service quality issues as the subject of negotiation or joint decision-making. As we have seen so far, this finding is common with many participation issues in this survey; in those two countries which enjoy the strongest form of participation, employee representatives' expectations are much higher. Allied to this observation is the fact that **whilst Danish managers are prepared to contemplate high levels of participation in the future to a level which largely meets the aspirations of their employee representative counterparts, German managers are not so willing to do so**.

As for the other countries, the differences between the levels of participation that management is prepared to concede in the future and the aspirations of employee representatives are not all that considerable except in the cases of **France** and **Italy**, where the differences seem to centre on whether or not product and service quality issues should be by information rather than consultation.

10.5 Assessing the results in the light of the Val Duchesse Social Dialogue

Throughout the various Joint Opinions which have emanated from the Val Duchesse process (see Chapter 3), particular emphasis is given to the need to "improve the competitiveness of European enterprises". In this respect the importance of product and service quality is central to the achievement of this aim. How do the countries rank in relation to the guide-lines laid down by the Joint Opinions?

If we consider the levels of participation that were in existence at the time the survey interviews were conducted, and take the responses of employee representatives as giving a critical view of such participation levels, we find that only in four Member States do a majority of respondents indicate that participation levels consist of either consultation or higher levels — **Greece, Ireland,** the **Netherlands** and the **United Kingdom.** There is a middle range of countries where around 30-40% of respondents indicate that the Val Duchesse bench-mark is being followed — **Belgium, Denmark, France, Germany,** and **Spain**. In three countries (**Luxembourg, Italy** and **Portugal**) less than 20% of respondents claim that the Val Duchess guide-lines are being followed.

What is the likelihood that the guide-lines emanating from the Val Duchesse process will be followed in the future? Here, it is better to consider the views of managers because it is managers after all who will be determining the parameters of participation in product and service quality in the future. The picture that emerges is much brighter; more than three out of four managers intend to follow the guide-lines in **Ireland, Denmark, Greece** and **the United Kingdom**. A majority of managers have the same intentions in **Spain, Belgium** and **France**. In four countries, the figure falls below 50% : **Germany** and **Italy** are still above 40% followed by **Luxembourg** and **Portugal** where only 14% of managers are ready to consult with employee representatives on quality matters.

10.6 Conclusions

Product and service quality have received particular emphasis throughout Europe in the 1980s; it is seen as essential if European enterprises are to compete in world markets during the years to come. As we have seen, its seeds were sown as a result of a sea-change in managerial philosophies during the 1980s and has been linked to notions of 'Total Quality Management' and new forms of 'Human Resource Management'. It is probably true to say that the most frequent form of participation in this area is a form of direct, task participation which has centred on the individual employee. For example, in one of the Member States with the highest levels of participation throughout our survey (**Germany**), there is extensive evidence that there has been a dramatic rise in the number of quality circles in German enterprises during the 1980s.[3] In some cases, such approaches from management have been designed to circumvent the influence of trade unions in the workplace. Since our survey covered enterprises with *formal* participation procedures with *indirect* forms of representation by the workforce, our results cannot measure the kind of task-centred participation which is centred on individual employees referred to above. Of course, it is always possible that when respondents were interviewed, they might well have referred to this form of employee-centred participation in their responses; we have no means of checking.

What kind of picture emerges from our survey results? It seems that at the time the the survey interviews were conducted, management in Europe tended to restrict employee representative involvement merely to the provision of information about product and service quality, and there were numerous examples of no involvement by employee representatives whatsoever. There was evidence of significant levels of consultation in those countries where most of the explanatory factors which were outlined in Chapter 4 were favourable, and there were some cases where this issue was subject to negotiation or co-determination.

Several countries stand out with favourable results in this area: **Denmark, Greece, Ireland,** the **Netherlands,** and the **United Kingdom.** The results for **Luxembourg** and **Portugal** were particularly poor. The traditional ideological reluctance of trade unions to be involved in what they see as being party to unpopular management decisions is also reflected in our survey results for **France** and **Italy;** however, there is also evidence that this is changing, with both management and employee representatives in those two countries expressing a willingness to see product and service quality as an issue where both sides can make a constructive contribution to the success of French and Italian enterprises.

Given the importance of this issue for the competitiveness of European enterprises in world markets, our results show that there is likely to be much stronger forms of participation in the years to come. In all countries there is a shift towards stronger forms of participation, but more so in some countries than others. In **Denmark, Greece, Ireland,** the

Netherlands, and the **United Kingdom** well over three out of every four managers indicate that future participation in product and service quality will be characterised by consultation and co-determination. The shift towards higher levels of participation is not so pronounced in **Belgium, Germany, Luxembourg** and **Portugal.**

One striking feature of our results is that whilst employee representatives in both **Denmark** and **Germany** are markedly enthusiastic about involving themselves in decisions about product and service quality, German managers, unlike their Danish counterparts, seem reluctant to move much further in their attitudes towards conceding higher levels of participation. This might suggest that German managers have reached their limit in the concessions they are prepared to make as far as collective workforce representational participation is concerned. Instead, German managers appear to be pursuing strategies of flexibility which emphasise more task-centred direct forms of participation focused on individual employees.

Generally, however, our survey results indicate that future participation levels in Europe in product and service quality are likely to be healthy. Both sides of industry appear to be aware of the importance of this issue for enhancing the competitiveness and productivity of European enterprises.

NOTES
1. R. Collard and B. Dale, in *Personnel Management in Britain,* K. Sisson (ed.), Blackwell, Oxford, 1989, p.371.
2. T. Peters and R. Waterman, *In Search of Excellence,* New York: Harper and Row, 1982.
3. See for example, Otto Jacobi and Walther Müller-Jentsch, in G.Baglioni and C.Crouch (eds.), *European Industrial Relations: the challenge of flexibility,* Sage Publications, London, 1990. According to these authors, the number of enterprises in Germany with quality circles during 1979-86 was as follows:
end of the 1970s   30
1982   200
1986   1,200-1,400

Chapter Eleven

Benefits of Participation

11.1 Introduction

The involvement of employees and their representatives in the process of technological modernisation is not an end in itself; it is an instrument by which the parties concerned expect outcomes which support their own interests. They might seek to avoid the adverse implications through participation or to achieve positive goals in the light of their own interests. It is natural that past experiences regarding the process of participation and its outcome shape the general climate and the opportunities for participation generally.[1] Any negative effects of participation will place a question mark on the whole idea, while visible benefits will create an atmosphere for the continuation and extension of a policy which has proven its worth.

11.2 Assessing the survey results

On the basis of the survey data available, we investigate the topic of perceived effects of participation in three ways:

— to give an overall impression on selected benefits of workforce representative participation;

— to compare the perception of management and employee representatives between the 12 Member States;

— to discover in what way the perceptions of both parties differ or are in line with each other;

As a starting point — and as a base for analysing these topics — we summarise briefly the experience both parties have had with the process of participation so far:

— As far as strategic topics are concerned, the more binding forms of involvement such as consultation, negotiation and joint decision-making were fairly rare.

— These higher levels of participation were more prominent in regard of direct workforce concerns as health and safety, work organisation and training.

— Both sides want more participation in the future. Workforce representatives are strongly involved in management's considerations and decisions about basic company problems.

Nevertheless, important differences remain. Management prefer for the future increased information and consultation, whereas employee representatives tend to seek negotiations and joint decision-making.

The overall impression which we gain of participation practices in high tech companies in the European Community is one of ambivalence: in strategic company matters the situation can be described as a combination of no involvement at all or low intensity of involvement of workforce representatives; whereas in operational problems there is a clear trend towards more intensive participation practices. Against this background we now analyse the issue of the effects of participation in three broad content dimensions:

— What is the impact of participation on the mutual understanding of the two parties in the firm? Does participation provide a base for better understanding between the two parties, or is it only a means for the institutionalised change within the intra-company power balance which might lead to increased conflict in the long run?

— Does participation improve the competitive position of the company by a better utilisation of skills or does participation have little influence on this strategic aspect of company policy?

— In what way does participation enhance or impede the process of technical modernisation of companies? This problem is analysed in regard to the practical issues of possible delay or acceleration in management decision-making in implementing technological change.

A first analysis of the perceived benefits of participation on an even wider range of indications came to the following conclusions:

"Reviewing the effects of participation in technical modernisation of European firms, the overall situation can be characterised as a mixture of no impacts and positive results. When managers and employee representatives indicated effects they were mainly, **sometimes overwhelmingly positive**. This is true even for the only sensitive issues of possible time delays through participation: decision-making and technology implementation. These are the aspects where managers as well as employee representatives report comparatively high rates of negative effects. But at the same time there are always more respondents who indicate that involvement procedures have helped to accelerate decision-making and technology implementation. In addition, there are comparatively high rates of both managers and workforce representatives who indicate an increased quality in decisions, and practically nobody voices negative repercussions in this respect.

This basic consent leads us to conclude that we are dealing with **largely co-operative, high-trust relationships between both sides in the European companies** under study. Both parties convey the message of rather **high consensus** between them. Such consensus carries the view of participation to work as a productive force to achieve common ends. The situation certainly does not point towards antagonistic relationships between management, the workforces and their representatives.

The results are more in keeping with a concept which stresses the positive aspects of technical innovation for the common good of all parties involved, where the benefits of various forms of participation are evident. We have discussed this approach in terms of positive-sum games, as situations of co-operation to where the pursuit of the interests of one side need not necessarily be detrimental to the other side. However, this does not mean that the conflict between the two sides will not emerge in the future".[2]

11.3 Mutual understanding in the firm

Good industrial relations generally are regarded as conducive to an optimal implementation and use of new technology. Within this part of the report we analyse the impact of participation on the attention paid by management to the concerns and interests of employees. This problem touches upon the social climate within the company as they deal with the question of empathy, i.e. the capability to perceive the problems and interests of the other side.

At the European level, managers and employee representatives express similar opinions about the effects of participation on this point. Both groups indicate a very low level of negative effects, respectively 1% and 5%.

Considering all respondents together, slightly more than one out of two indicate that workforce involvement has not changed the situation. Among managers, the same proportion estimates either that the situation is unchanged or that the situation has improved. The employee representatives are a little more pessimistic; 40% of them consider that workforce involvement has had positive effects while 55% indicate that participation has had no effect.

If we turn to the answers given in each of the 12 Member States, the most important point to notice is the great similarity across countries. Indeed, the percentage of positive effects acknowledged by managers vary between 43% (Luxembourg) and 57% (UK).

Yet, if we consider the responses of managers, it should be noticed that it is in **Denmark** that we find the highest proportion of negative effects, which is still as low as 6%, and one of the lowest rates of unchanged situations. This situation could partly be explained by the high level of participation in **Denmark** which might have heightened the expectation level of managers.

BENEFITS OF PARTICIPATION

Figure 54: Effects of participation: Attention by management to the concerns and interests of employees

- Managers
- Employee represent.

deteriorated: 1% / 5%
unchanged: 49% / 55%
improved: 50% / 40%

Source: Survey in all EC Member States, 1987-1988; 3 848 Managers and 3 848 Employee Representatives.

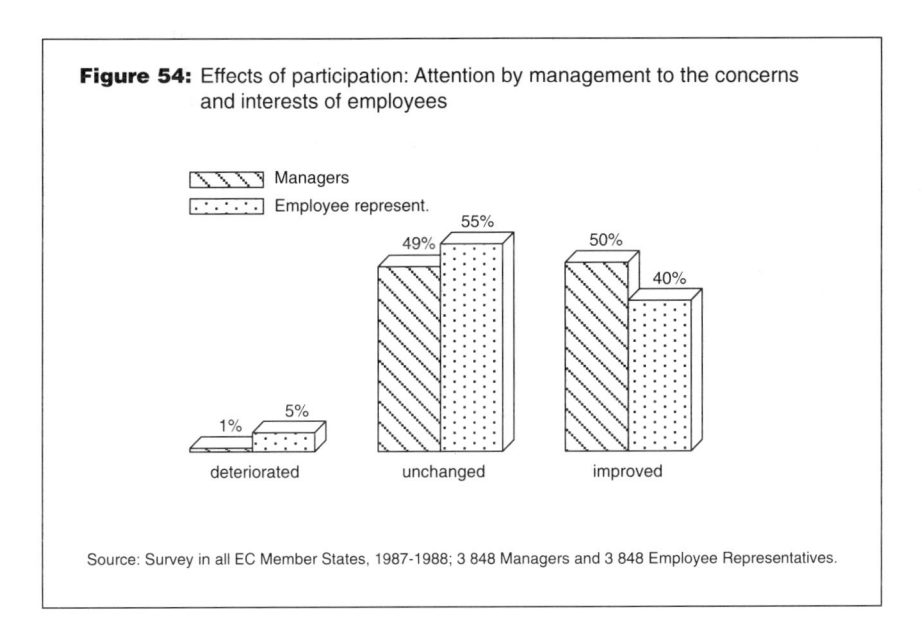

In **Portugal,** we find the highest level of an unchanged situation: nine out of ten managers indicate that workforce involvement has had no effect on the attention they give to workforce interests. This result should be correlated with the fact that participation is a very recent procedure there.

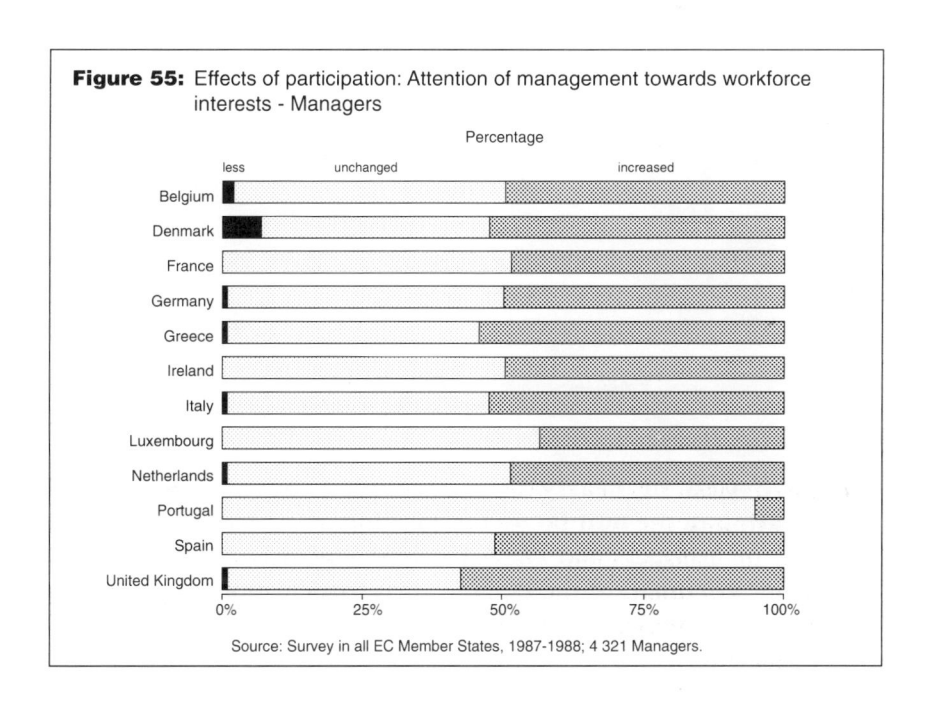

Figure 55: Effects of participation: Attention of management towards workforce interests - Managers

Percentage — less / unchanged / increased

Belgium, Denmark, France, Germany, Greece, Ireland, Italy, Luxembourg, Netherlands, Portugal, Spain, United Kingdom

Source: Survey in all EC Member States, 1987-1988; 4 321 Managers.

171

WORKPLACE INVOLVEMENT IN TECHNOLOGICAL INNOVATION IN THE EC

British managers turn out to be the most positive; nearly six out of ten indicate that their attention towards workforce interests has been increased. This more positive tendency might be related with the absence, in the United Kingdom, of a formal institutionalised system of industrial democracy at company level. Participation linked with the introduction of technological change might have offered new opportunities. But it should not be forgotten that four managers out of ten still indicate that participation has had no effect.

If we consider the point of view expressed by employee representatives, the pattern of answers varies little between countries. The percentage of respondents indicating increased attention by managers varies between 37% in the **Netherlands** to 51% in **Ireland**.

The difference between managers and employee representatives concerning positive effects is highest in the **Netherlands** where it does not exceed 13 percentage points.

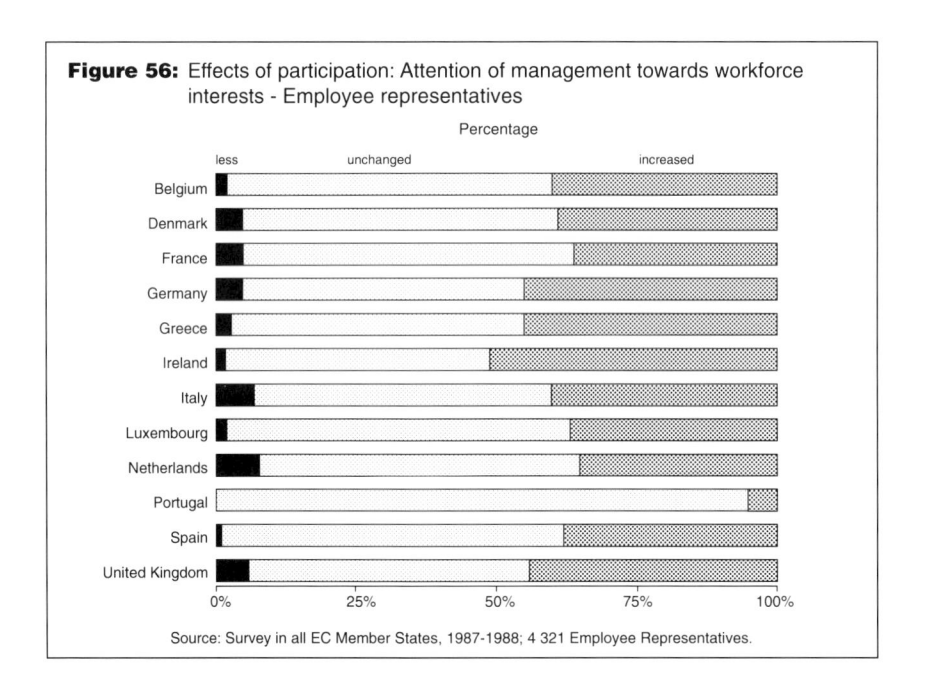

Figure 56: Effects of participation: Attention of management towards workforce interests - Employee representatives

Source: Survey in all EC Member States, 1987-1988; 4 321 Employee Representatives.

Thus, we find in nearly all countries a positive impact of participation in the mutual understanding in the firm. Between 35 and 55% of managers and employee representatives in 11 countries (exception: **Portugal**) of the EC agree that participation has improved the social climate between the two sides of industry. The negative impacts are negligible.

11.4 Utilisation of knowledge and skills

The skills and knowledge of the workforce are potential bottle-necks in the process of technical modernisation of companies. The lack of skills can slow down the modernisation process. Frequently, the skill level available is an important factor in planning for new technology: complex technologies are less likely to be introduced where the skill level of the workforce is low.[3] In addition, existing skills have to be mobilised in order to make the introduction of new information technology a success. In this context, the involvement of workforce representatives has had a very positive impact.

If we now consider the effects of participation on the utilisation of knowledge and skills, European managers and employee representatives have given practically the same answers. They both estimate the negative effects of participation on the utilisation of knowledge and skills of employees. Nearly 45% indicate that participation has not changed the situation.

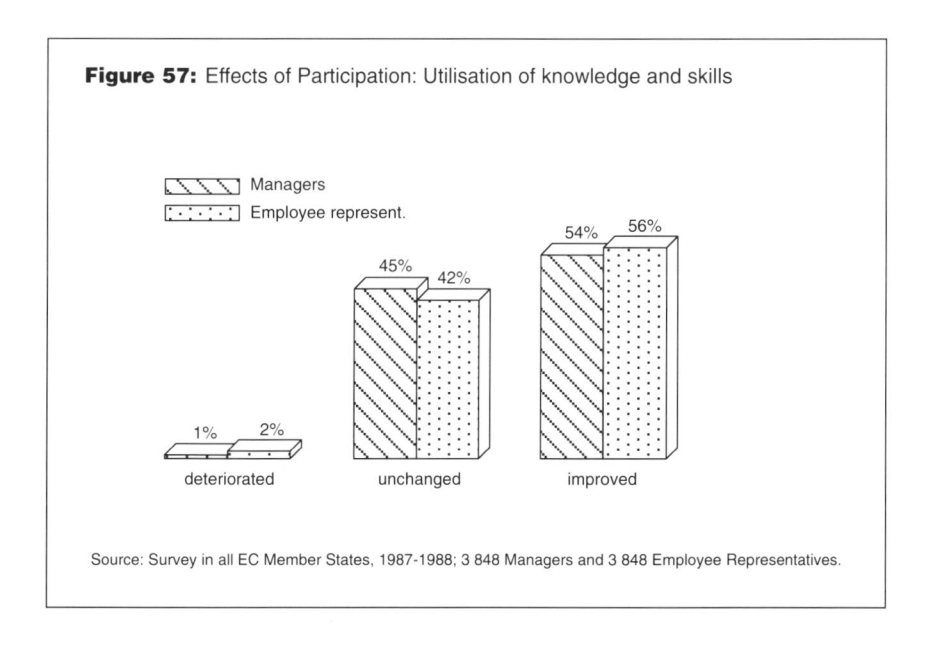

Figure 57: Effects of Participation: Utilisation of knowledge and skills

Managers: deteriorated 1%, unchanged 45%, improved 54%
Employee represent.: deteriorated 2%, unchanged 42%, improved 56%

Source: Survey in all EC Member States, 1987-1988; 3 848 Managers and 3 848 Employee Representatives.

At country level, we also observe that in the majority of countries, positive effects are more frequently acknowledged by managers than an unchanged situation. **Ireland, United Kingdom** and **Denmark** are the countries where most managers (70% and more) report positive effects of workforce involvement. Further down the figure, in decreasing order, **Greece, Netherlands, Belgium** and **Italy** have high percentages of positive effects.

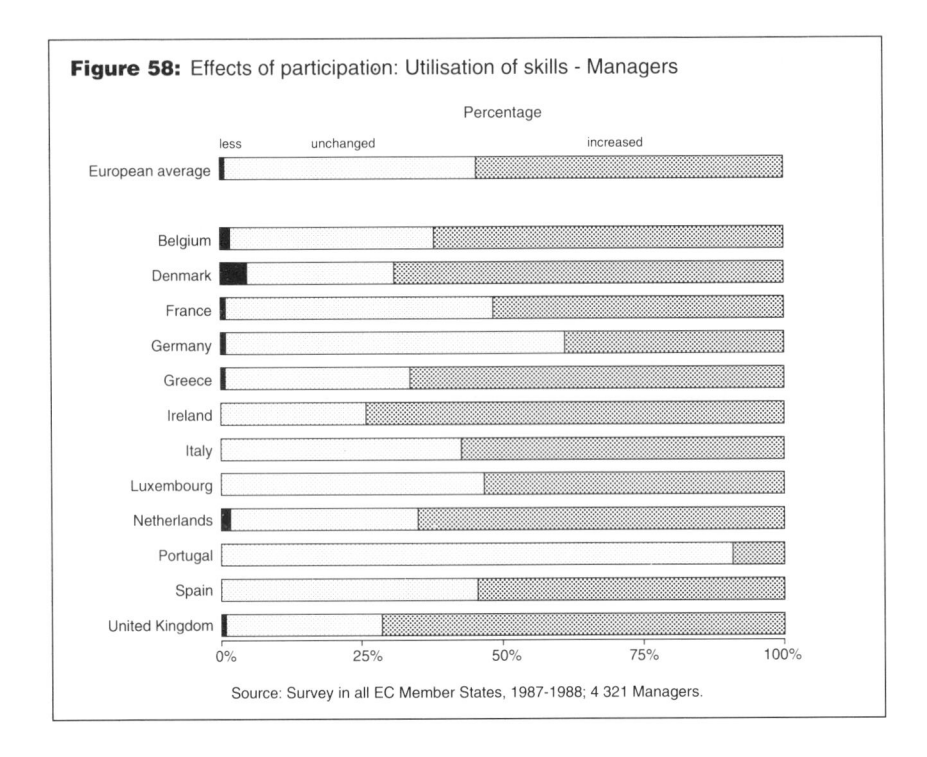

Figure 58: Effects of participation: Utilisation of skills - Managers

[Horizontal stacked bar chart showing percentages (less / unchanged / increased) for European average, Belgium, Denmark, France, Germany, Greece, Ireland, Italy, Luxembourg, Netherlands, Portugal, Spain, United Kingdom, on a 0%–100% scale.]

Source: Survey in all EC Member States, 1987-1988; 4 321 Managers.

Among the Member States where managers report that participation has had little effect, **Portugal** must again be set aside with its large proportion of 91% of firms in which participation has had no change. The situation in **Germany** is more remarkable. Although it is usually characterised by a high level of participation, it is ranked after **Portugal** with 61% of managers reporting that workforce involvement did not increase the utilisation of skills and knowledge of employees.

If we compare the responses of employee representatives with those of managers, they appear similar in most countries. For ten countries out of twelve the difference between both sides of industry does not exceed 15 percentage points. The largest difference is to be found in **Luxembourg** where it reaches 26 percentage points. The second largest is to be found in **Germany** where 61 percentage points of employee representatives believe that participation has improved the utilisation of skills, i.e. 23% more than their manager counterparts. This difference in the perception of management and employee representatives in **Germany** might indicate that the interrelation between participation and qualification and skills is a matter of controversy in German companies.

In general we can conclude that from the point of view of the two sides of industry in most European countries **participation leads to a better**

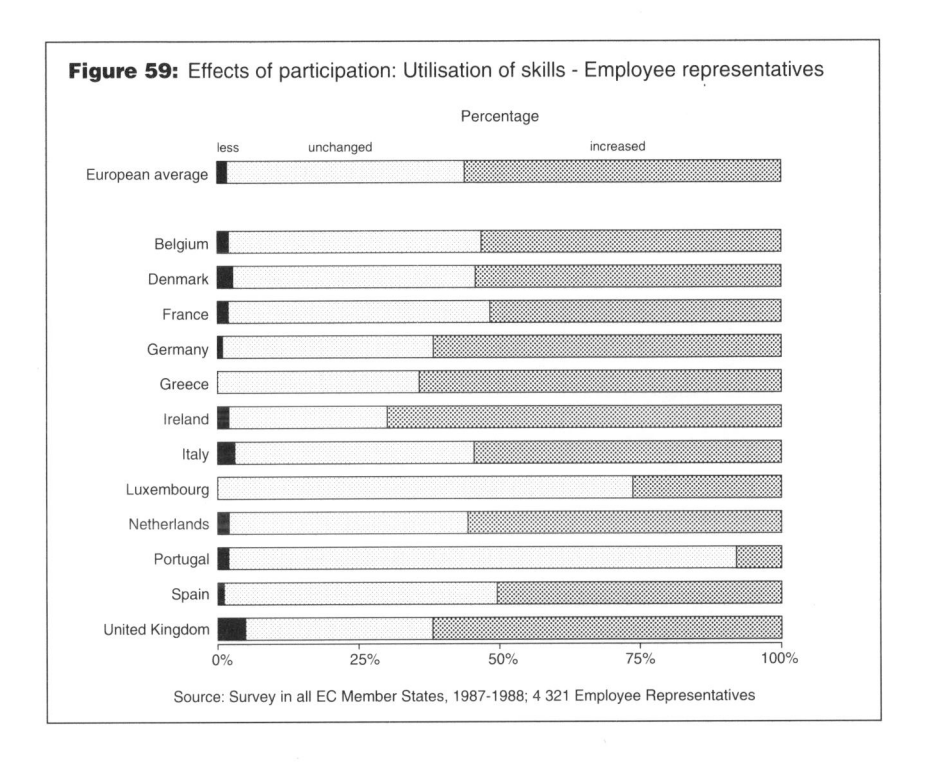

Benefits of Participation

Figure 59: Effects of participation: Utilisation of skills - Employee representatives

Source: Survey in all EC Member States, 1987-1988; 4 321 Employee Representatives

utilisation of knowledge and skills in the process of technical modernisation of companies. However participation alone is ineffective in situations where there is a lack of basic professional qualifications, this lack of skill qualifications is one of the key problems in many peripheral areas of the EC.

11.5 Decision-making

One of the central arguments against participation in technological change from a management point of view is the fear that company performance may be adversely affected by prolonging the time needed for decision-making or by a general decrease in the quality of decision-making. This expectation can be rejected on the basis of results of our study. Nearly a quarter of manager and worker representatives in Europe perceive a reduction in the time needed for decision-making and two thirds report that it has no effect at all. Only one out of ten managers (11%) in European high-tech companies reported that participatory procedures prolonged decision-making.

Figure 60: Effects of participation: Time needed to arrive at decisions

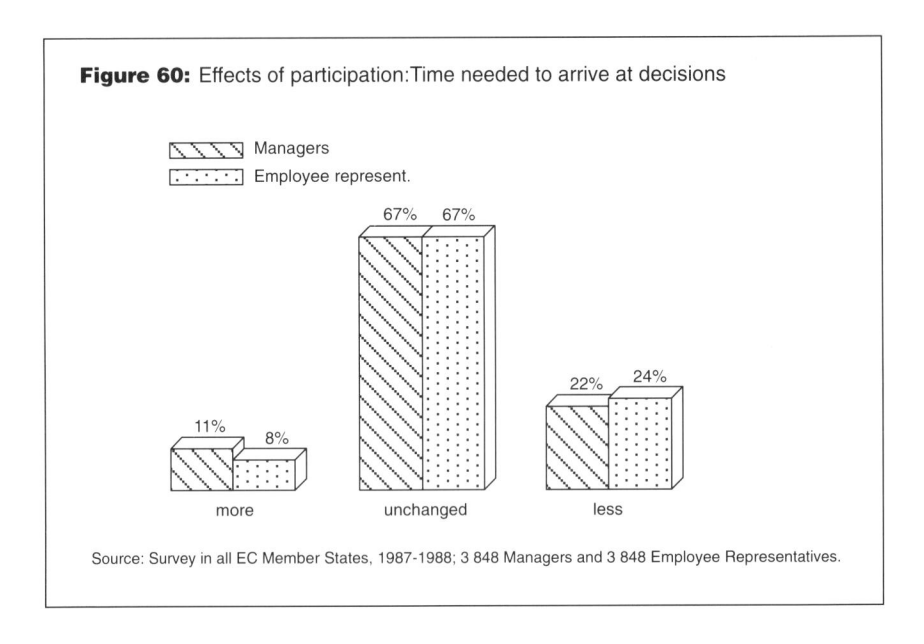

- Managers
- Employee represent.

	more	unchanged	less
Managers	11%	67%	22%
Employee represent.	8%	67%	24%

Source: Survey in all EC Member States, 1987-1988; 3 848 Managers and 3 848 Employee Representatives.

Figure 61: Effects of Participation: Quality of decisions

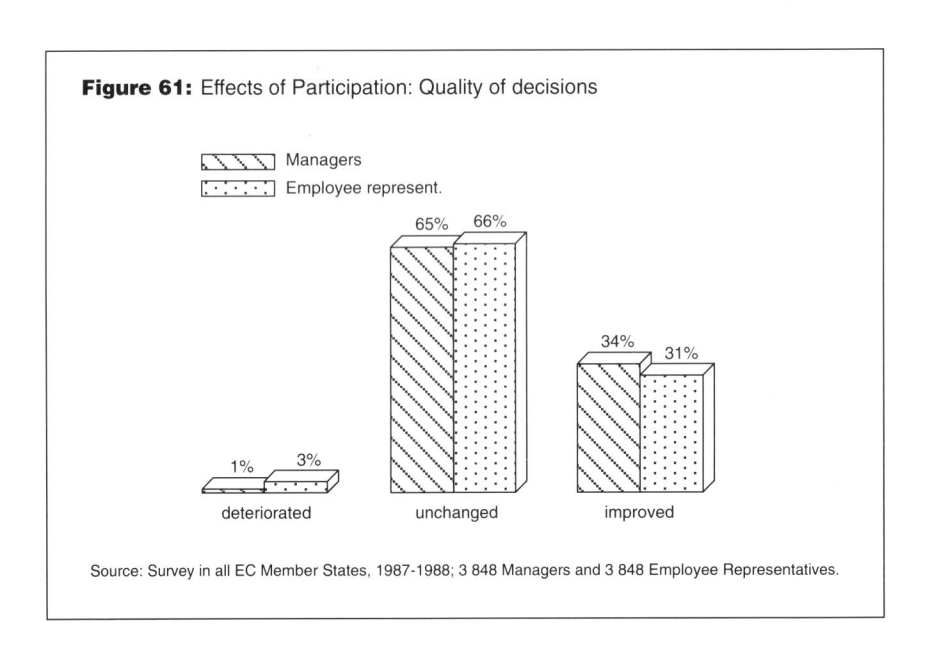

- Managers
- Employee represent.

	deteriorated	unchanged	improved
Managers	1%	65%	34%
Employee represent.	3%	66%	31%

Source: Survey in all EC Member States, 1987-1988; 3 848 Managers and 3 848 Employee Representatives.

The results are even more positive if one analyses the impact on the quality of decision-making. One third of management and employee representatives report an improvement in the quality of decision-making; two thirds on both sides perceive no impact. Negative effects are negligible (1 and 3%).

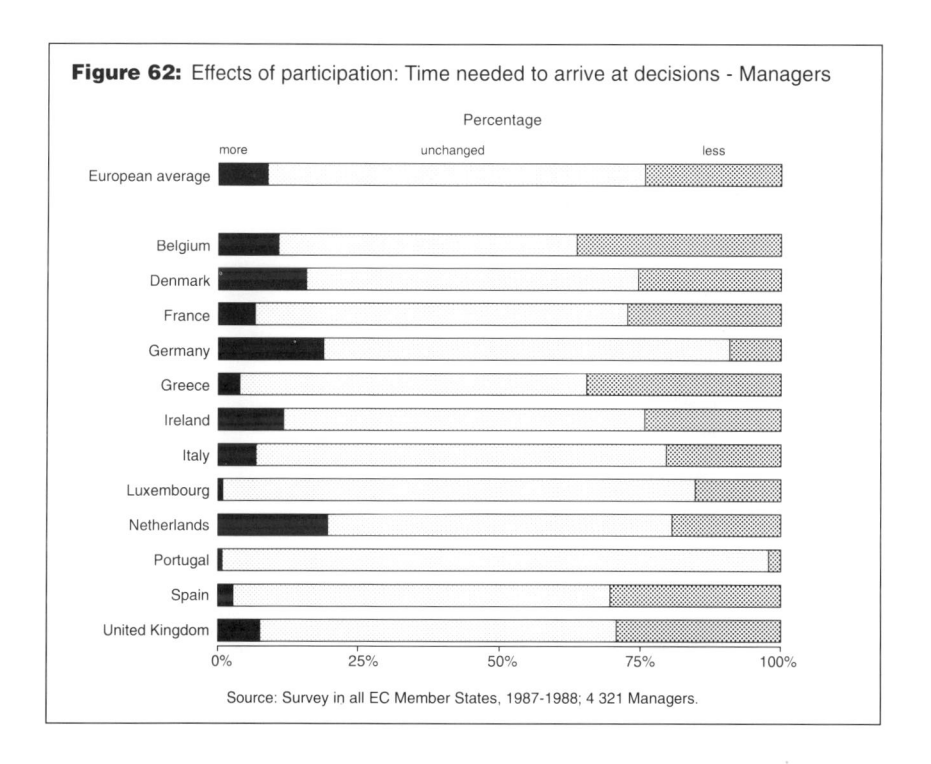

BENEFITS OF PARTICIPATION

Figure 62: Effects of participation: Time needed to arrive at decisions - Managers

Source: Survey in all EC Member States, 1987-1988; 4 321 Managers.

What is the impact on the time needed for decision-making in the 12 Member States? In each of the Member States, we find a majority of respondents who think that participation had no effect on the amount of time needed to arrive at decisions. The percentage of managers sharing that opinion varies between 53% in **Belgium** and 84% in **Luxembourg** — except for **Portugal** where it reaches 97%. There are also variations in the importance of positive and negative effects in the European firms but they are not so wide.

If we compare positive and negative effects, we can observe that in the majority of countries, positive ones surpass the negative ones. It is in **Belgium, Greece, Spain** the **United Kingdom** and **France** that participation had the most positive effects.

In the **Netherlands**, on the other hand, the number of firms where positive effects are reported is equal to the number of firms reporting negative effects. And in **Germany**, the number of **managers considering that workforce involvement has slowed down decision-making is twice as high as the number of managers who see an improvement in a time perspective**. These results are worth noting as these two countries have a high level of participation. Moreover, in **Germany**, our survey results generally show differences of opinion between management and employee representatives.

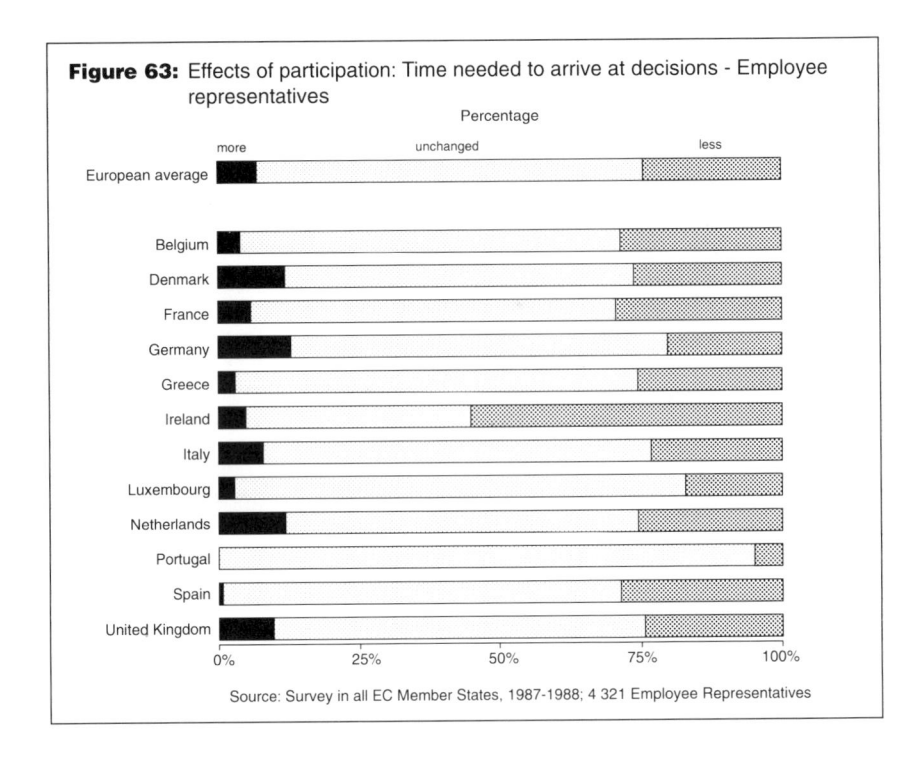

Figure 63: Effects of participation: Time needed to arrive at decisions - Employee representatives

Source: Survey in all EC Member States, 1987-1988; 4 321 Employee Representatives

The pattern of employee representatives' answers shows even more interesting variations. In five countries, we find significantly more positive perceptions as negative. In **Ireland** for 55% of employee representatives the time needed has decreased and only 5% see an increase. A similar pattern, but less pronounced, can be found in **Belgium, France, Greece** and **Spain**. Apart from the Portuguese, the most sceptical perception is given by German employee representatives. One in five perceive a positive but 13% a negative impact of participation on the time needed for decision-making, i.e. the reservations of German managers are partly shared by the other side.

The second dimension in regard to decision-making was related to the impact of participation on the **quality of decision-making**. Here, we observe significant differences between the managers of the 12 Member States. The most positive assessment is given by Danish management; 56% of them report that the quality of decision-making has improved. Also in the **Netherlands** (48%), in **Greece** (47%), the **UK** (45%) and **Ireland** (42%) one finds an improved quality of decision-making, which is

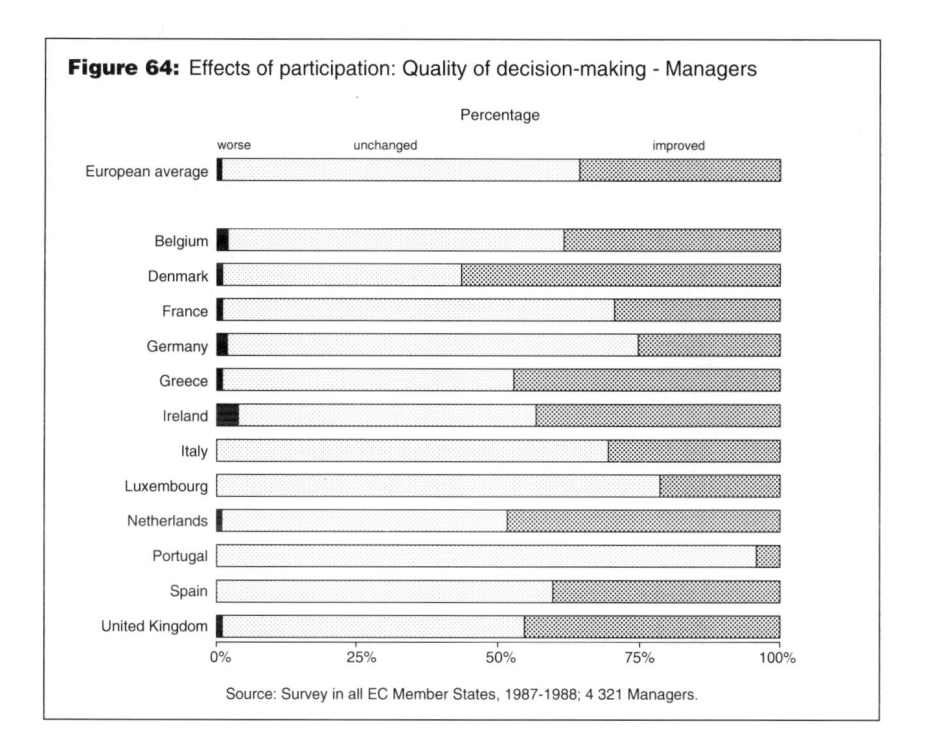

Figure 64: Effects of participation: Quality of decision-making - Managers

Source: Survey in all EC Member States, 1987-1988; 4 321 Managers.

significantly higher than the European average. Apart from **Portugal**, the lowest level of improvements is reported by managers in **Luxembourg** (21%), **Germany** (25%) and **France** (29%).

On the whole, the pattern of employee representatives' answers is similar. Everywhere, we find a majority of respondents who state that participation had no effect and very few of them think is has worsened the quality of decision-making. In most countries, workers' representatives are close to managers in their view of the positive effects of participation some see somewhat less improvement, this difference being greatest in the **United Kingdom** (15 percentage points). The only country where more employee representatives than managers believe that participation has improved the situation is **Germany**, but the difference is only of 12 percentage points.

Thus, there is convergence between both sides of industry on this point with a majority of respondents seeing no effect of participation.

As a conclusion one can assume that there is a high level of agreement between both sides of industry in Europe on the impact of participation on decision-making. **Gloomy expectations** voiced by some politicians, officials in employers' organisations and in the management literature **can**

Figure 65: Effects of participation: Quality of decision-making - Employee representatives

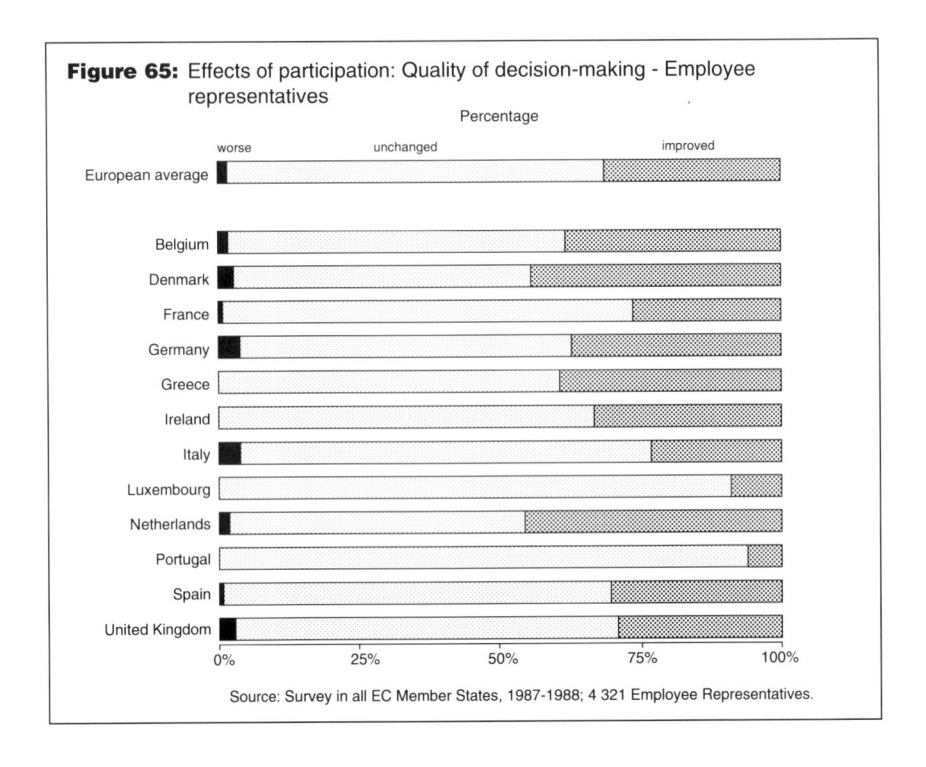

Source: Survey in all EC Member States, 1987-1988; 4 321 Employee Representatives.

Figure 66: Effects of Participation: Time needed for technology implementation

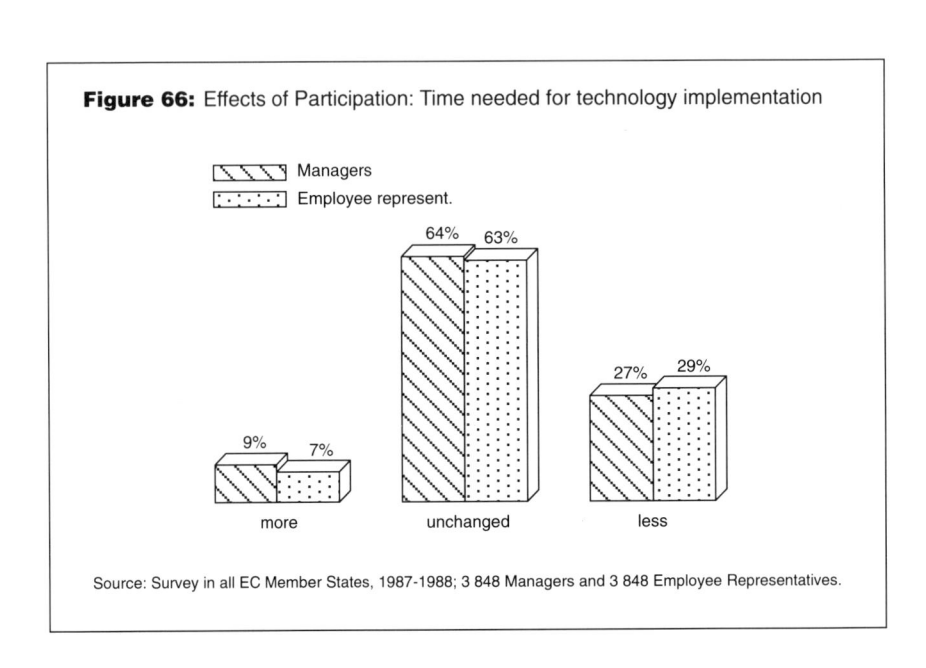

- Managers
- Employee represent.

	more	unchanged	less
Managers	9%	64%	27%
Employee represent.	7%	63%	29%

Source: Survey in all EC Member States, 1987-1988; 3 848 Managers and 3 848 Employee Representatives.

Benefits of Participation

be rejected by the perception of management of a neutral or positive impact of participation. **The most sceptical management can be found in Germany**. Its perception differs significantly from the positive assessment of Danish and Dutch management, which have either a similar practice of participation or similar regulations.

11.6 Implementation of new information technology

The competitiveness of companies can be strongly affected by how long it takes to introduce new technology equipment and systems. To what extent does participation of employee representatives influence the time needed for the implementation of new technology? Generally, we observe at the European level similar results in comparison with the time needed for decision-making. There are around two thirds of both types of respondents who indicate no effects on participation but around 8% find that workers' involvement has increased the time needed for technology implementation. Positive effects clearly outweigh the negative ones.

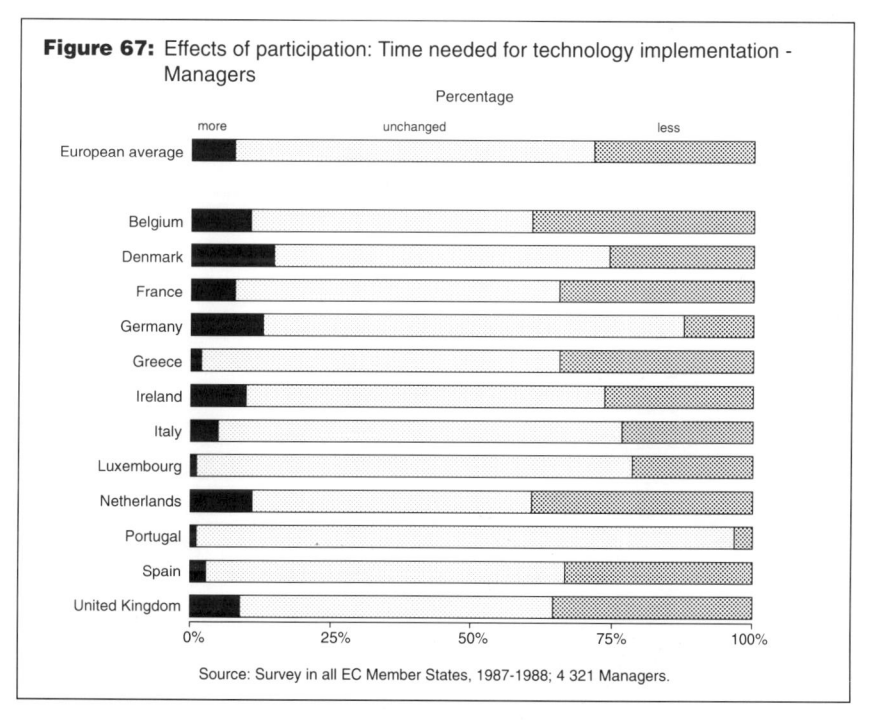

Figure 67: Effects of participation: Time needed for technology implementation - Managers

Source: Survey in all EC Member States, 1987-1988; 4 321 Managers.

At country level, the percentage of managers reporting no change of situation as a consequence of participation varies between 50% and 78% with the exception of **Portugal** where it is 96%. Just above **Portugal** in the figure, we find **Luxembourg, Germany** and **Italy** where three respondents out of four report no effects.

The proportion of managers considering that participation has positive effect varies between 12% (**Germany**) and 39% (**Belgium**) while the proportion of those considering that participation has negative effects, varies between 1% and 15% (**Denmark**).

If we compare now the differences between positive and negative effects, we observe that in nine countries, those reporting positive effects exceed those reporting negative effects by at least 17 percentage points. The difference between positive and negative effects is negligible in **Germany**, but there is the largest proportion of managers who report an unchanged situation. It is in Danish firms that we find the highest proportion of negative effects: 15%. So there might be some link between the level of participation and the level of expectation which can diminish the proportion of those who report positive effects. Yet the **Netherlands** is among the countries where positive effects are most frequently reported, i.e. by more than 30% of managers, the other ones being **Belgium, France, Greece, United Kingdom** and **Spain**.

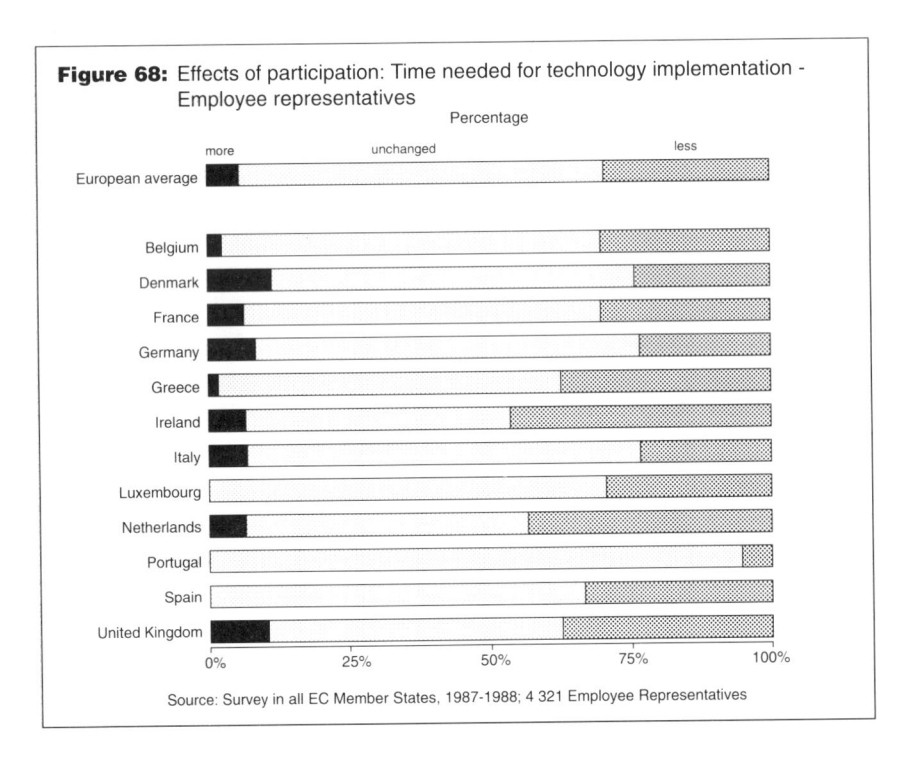

Figure 68: Effects of participation: Time needed for technology implementation - Employee representatives

Source: Survey in all EC Member States, 1987-1988; 4 321 Employee Representatives

There are significant variations in the perception of employee representatives. Nearly half of Irish and Dutch employee representatives observe a strong reduction in the time needed for implementation. The other extreme can be found in **Denmark, Germany** and **Italy**. Only a quarter of the respondents in these three countries see a positive impact of participation in implementing new technology.

11.7 Conclusions

Reviewing the effects of participation in the technical modernisation of European firms, the **overall situation** can be characterised as a **mixture of no impacts and positive results**. When managers and employee representatives indicated effects they were mainly — sometimes overwhelmingly — positive. This is true even for the sensitive issues of possible time delays through participation: decision-making and technology implementation. These are the areas where managers as well as employee representatives report comparatively high rates of negative effects. But at the same time there are always more respondents who indicate that involvement procedures have helped to accelerate decision-making and technology implementation. In addition, there are comparatively high rates of both managers and workforce representatives who indicate an increased quality in decisions, and practically nobody voices negative repercussions in this respect.

Further, the mutual understanding of both management and the workforce for each others' problems and constraints have been improved under the impact of participation in many countries. We have reason to believe that these positive developments are embedded in a context of predominantly high trust relationships which existed prior to the experience of participation in technical change, but which have since forged a more improved co-operative climate. The majority of both sides indicate a better utilisation of workforce skills due to participation. Knowing that new technology is best applied under conditions of a highly skilled workforce, these results again point to conditions in the companies which were conductive to participation to begin with.

A comparison between the 12 Member States shows important differences in the assessment of the benefits of participation. Portuguese managers are least positive in all five dimensions. Their general perspective is that participation has no impact on the economic and social performance of the company. This general attitude is also taken by the Portuguese employee representatives. **It is important to note that in regard to the benefits of participation no joint pattern for the peripheral countries of the EC can be observed. Portugal** is the exception. The reason for this may be the immaturity of the Portuguese system of industrial relations, the short democratic tradition and the recently introduced legislation in this area. But these arguments are not fully convincing as they **also** partly apply to **Spain** and **Greece**.

An even more astonishing feature is the negative attitude of German managers. In regard to the key criteria for the valuation of participation, German management is significantly more pessimistic than the European average on the management side. The German results are even more exceptional, if one compares them with the Danish, Dutch and Belgian results. **Denmark** and **Germany** are the most advanced countries with regard to participatory practice in the EC. Within Danish management two out of five categories are significantly more positive than the European average. From a previous study, we know that there is a statistically significant relationship between the strength of participation and its positive evaluation; but this is **not** true for German management. The reason for this may be tactical reasoning, i.e. that a positive response may be used against German management to reinforce legislation.

One also can compare the attitudes of Danish and German managers on another level. Previous research showed that Danish managers were much more open to strong forms of participation in strategic areas for the future than German managers. It seems to be that Danish managers, due to their positive experience with participation, are much more convinced about its value, whereas German management seems to be pushed into participation by legal regulation and strong unions.

Another interesting observation is that there is no relation between strong legal regulations and negative management perception of benefits. **Belgium** and the **Netherlands** have strong legal regulations on one side, but management is significantly more positive than the European average in one (**Netherlands**) or two (**Belgium**) dimensions.

The last interesting result is related to the extraordinary positive assessment of British management in all five dimensions. An explanation may be the selection of UK companies. The survey sample consisted of companies with formal employee representation and some minimal practice in participation, i.e. considering the hostile political environment for participation in 1987, the companies in the survey were those which were prepared to develop participation without legal or contractual obligation and against a hostile political environment. It is not unreasonable to assume that it needs a strong conviction of management based on positive impacts of participation to develop some participatory practice under these circumstances.

The enthusiasm of British management **is not shown by British employee representatives**. Their perception does not differ from their

European colleagues with the exception of one dimension. The most negative results can be found in **Portugal** and **Luxembourg**. The perceptions of employee representatives in **Luxembourg** is in line with those of German management. A system of co-operative industrial relations produces doubts on the benefit of participation. The most positive employee representatives in Europe can be found in **Ireland**. This is another indication that Irish industrial relations is increasingly diverging from the British tradition. Irish workers' representatives and unions are much more positive about participation than their UK counterparts. On the management side it is vice versa. The German and Danish workers' representatives show only in one dimension more positives attitudes than their European colleagues.

Summing up the country comparison one is confronted with a complex pattern not only between the 12 Member States but also between management and workers representatives in the Member States.

Tabular summaries of significant deviations of particular countries on various dimensions of benefits of participation in technological change from the European average by managers (**TABLE 5** below) and employee representatives (**TABLE 6** next page).

TABLE 5: MANAGERS

Country	Mutual understanding	Utilisation of knowledge & skill	Time needed for decision making	Quality of decision making	Time needed for implementation
Belgium			(+)		(+)
Denmark		(+)		(+)	
France					
Germany		(—)	(—)	(—)	(—)
Greece			(+)	(+)	
Ireland		(+)			
Italy					
Luxembourg				(—)	
Netherlands					(+)
Portugal	(—)	(—)	(—)	(—)	(—)
Spain			(+)		
UK	(+)	(+)	(+)	(+)	(+)
Dimension	"better"	"better"	"less"	"better"	"less"

(+): significantly "better" or "less" than European average
(—): significantly below the European average.

TABLE 6: EMPLOYEE REPRESENTATIVES

Country	Mutual understanding	Utilisation of knowledge & skill	Time needed for decision making	Quality of decision making	Time needed for implementation
Belgium					
Denmark				(+)	
France					
Germany		(+)			
Greece		(+)			
Ireland	(+)	(+)	(+)		(+)
Italy					
Luxembourg		(—)	(—)	(—)	(—)
Netherlands				(+)	(+)
Portugal	(—)	(—)	(—)	(—)	
Spain					
UK		(+)			
Dimension	"better"	"better"	"less"	"better"	"less"

(+): significantly "better" or "less" than European average
(—): significantly below the European average.

NOTES
1. A comprehensive discussion of this question can be found in D. Fröhlich, C. Gill, H. Krieger, *Roads to Participation,* Office for Official Publication of the European Communities, Luxembourg, 1991.
2. D. Fröhlich, C. Gill, H. Krieger, *ibid.* 1991, pp. 100-101.
3. A. Sorge et al: *Microelectronics and Manpower in Manufacturing: Applications of CNC in Great Britain and West Germany,* London, 1983.

Chapter Twelve

Conclusions

This report has dealt with a very wide range of issues concerned with participation by employee representatives in the introduction of new information technology into enterprises throughout the European Community. As such it presents a fairly comprehensive picture of the various degrees of participation that were in existence at the time the interviews were conducted in each Member State in various aspects of technological change: participation in ameliorating the adverse effects of technological change on individual employees; participation in issues concerned with training; health and safety; work organisation; investment in new information technology and in product and service quality. It also provides an indication of the likely degrees of participation in the future and identifies the importance attached to particular aspects of technological change by managers and employee representatives throughout the Community.

This is the **only** survey which covers participation in technological change throughout the EC and therefore its results are particularly important to all the Social Partners at a time when the Community is moving towards greater economic and political union and when various social policy initiatives are being proposed involving information and consultation of employees.

What are the major implications for the Social Partners which emerge from this survey? Perhaps the most important finding is that where participation occurs, regardless of the issues involved, it is overwhelmingly seen by both sides as **beneficial and positive**. It is seen as conducive towards promoting better mutual understanding between management and employees in companies, a greater utilisation of knowledge and skills of employees in the process of modernising companies, a smoother process of technological implementation, and either a neutral or positive effect on the quality of decision-making. In general, our results point to a high degree of trust between both sides of industry.

Another important conclusion is that there is evidence that managers and employee representatives are both seeking higher levels of participation in technological change in the immediate future. Both sides value the contribution that can be made by each other to promoting a smoother transition towards the modernisation of the enterprises in which they work. While there are sometimes differences in the degrees of participation that the two sides would prefer in the future, variations in the preferences as between different aspects of technical change, differences across the 12 Member States, there is nevertheless **a clear trend towards improved co-operation between the two sides and higher degrees of participation in the future.**

Given the high degree of consensus which we found in our results, it is worth also stressing that all our findings suggest that employee representatives — many of whom were trade unionists — are unreservedly enthusiastic about technological change. Indeed, our findings are entirely consistent with the work of Daniel[1] in the U.K. This was the case across all the Member States of the Community and in manufacturing as well as in the office environment. There was no evidence of widespread employee representative resistance to technological change; indeed, our findings reflect the position taken by the ETUC in its discussions with UNICE and CEEP (see Chapter 3) in producing a number of Joint Opinions concerned with the introduction of new technology.

Earlier in this Report we mentioned that our results should be seen in the light of a number of political initiatives that are being taken to promote participation in European enterprises as part of the Social Dimension of the internal market in 1992. One important (and often underrated) strand of European social policy is the Val Duchesse Social Dialogue. Our results were assessed against the various Joint Opinions which have emanated from this process. It is particularly significant that a recent understanding between UNICE and the ETUC has enabled Commission proposals on the Social Action programme to implement the Social Charter to be considered within the Val Duchesse process to see whether or not both sides can produce an agreement between themselves within a nine-month period before the Commission needs to issue a Directive. Our results can act as a benchmark in their deliberations.

When we consider the levels of past and future participation across the range of issues concerned with technological change, it is clear that there are variations in the degree of participation from one issue to another. When we compare the levels of participation against each other across this range of issues, we find that they are highest in health and safety and relatively high in work organisation and in training. They are lowest in product and service quality and in investment strategy.

However, this comparison between the participation levels can be misleading taken on its own. Even though participation levels in **health and safety** are the highest of all the issues of technological change which we explored, in four of the Member States (**Spain, Portugal, Luxembourg** and **France**) less than half of all managers report that there is any consultation between the two sides. Given the importance of health and safety as an issue which has received a great deal of attention in the Community, such results **are a matter for concern.**

The same could be said about participation levels in another issue which has been been singled out as being particularly important during the 1990s

— that of **training**. Here, there are six Member States (**Spain, Portugal, Luxembourg, Italy, Greece** and **Germany**) where over half of all managers report that there is not any consultation between the two sides, although in **Germany** there is admittedly a comparatively high degree of negotiation and joint decision-making.

As for **work organisation**, over half of managers in five Member States report that there is either no involvement by employee representatives or that participation is characterised by the mere provision of information (**Spain, Portugal, Luxembourg, Italy,** and **France**). As we stressed in Chapter 8, this is an issue which lies at the centre of employee concerns about the need to participate in the jobs affecting their members; indeed, these results are worryingly low because if participation means anything it has to cover this area.

Perhaps the most worrying of all issues as far as participation levels are concerned in relation to the interests of management is that of **improving product and service quality**. Here, there are seven countries where over half of all managers state that either there is no involvement whatsoever by employee representatives or such matters are the subject of the mere provision of information (**Spain, Portugal,** the **Netherlands, Luxembourg, Italy, Germany** and **France**). Given the importance of this issue in contributing to the competitiveness of European enterprises in world markets, such results are worrying indeed.

The degree of participation in **investment strategy** is lowest of all; that, however, is only to be expected. Nevertheless, two countries — **Denmark** and **Germany** have the highest degree of participation in this aspect of technological change.

The picture that emerges from a country comparison of participation levels in the aspects of technological change that have been explored in the survey show a distinct **North-South** divide among the Member States in the Community. Northern European countries such as **Denmark, Germany,** the **Netherlands, Belgium** the **United Kingdom** and **Ireland** all have higher levels of participation than those in the South. It is clear that participation is underdeveloped in the southern Member States — **Portugal, Spain, Italy, France, Greece** (with the inclusion of **Luxembourg** in this group).

How can participation be improved in the southern Member States? If participation is to develop further in these Member States then some form of **institutional support** is required in order to create a favourable climate in which the seeds of participation can grow. How can such favourable conditions be created? In order to answer this question we must return to Chapter 4 of this Report where we used an explanatory framework as a means of accounting for differences in participation levels between Member States. Our explanatory factors which were outlined in Chapter 4 proved particularly useful in interpreting our results and explaining the differences in participation levels between the various Member States. The North-South divide referred to above can be readily accounted for by using our explanatory framework. The five factors can also be used as a means of identifying the conditions whereby participation can be improved in these southern countries.

If we take each factor in turn it provides some clues as to the kind of institutional support for participation that might emerge in the future. The technological objectives which are pursued by management when introducing new information technology clearly influences their dependence on the skills and problem-solving capacities of their employees. As technological change becomes more sophisticated and increasingly concerned with product as opposed to process innovation in enterprises in these southern countries, so too can we expect more favourable conditions for participation to develop.

Insofar as management style is concerned, there is every indication that the quality of management generally is improving and we can expect that they will become increasingly aware of the advantages that participation can bring to improving productivity and competitiveness in European enterprises. In time, this will have a favourable effect on management style and attitudes throughout the Community.

Many of the obstacles that face trade unions in these southern countries in seeking greater influence in the enterprises in which their members work are caused by religious and/or political divisions in the trade union movements or a lack of a well-trained cohort of representatives who are able to negotiate with management. In some countries, as we saw earlier, there is also a legacy of dictatorial régimes which have smothered trade union development. However, this is all changing for the better, as rival union federations are increasingly co-operating in their dealings with governments and management (e.g. **Spain, Portugal, France** and **Italy**) and adopting a more flexible posture in bargaining. We can therefore expect improvements here in the future.

The question of regulation, particularly legislation, is not as simple as it might appear to be at first sight. Many mistakenly believe that participation can be created simply by introducing legislation or Community-wide Directives. However, while legislation **may** be the answer in some cases, it does not necessarily follow that the introduction of such legislation can **guarantee** that participation will occur in practice. There is ample evidence from our survey — particularly on health and safety — to suggest that the mere existence of legislation is no guarantor of participation. All this is to emphasise that **on its own** legislation is not necessarily the answer if we are to see improved levels of participation generally; it is merely **one** of a number of factors that are important. Participation depends on the willingness of both sides of industry in order to make it work. However, this is not to say that legislation supporting participation is always unnecessary; in some cases it it will be essential.

The degree of centralisation of the industrial relations system will probably work against the development of participation in the future. Throughout Europe there is a general trend towards more decentralisation in bargaining and in management decision-making; perhaps the omens here are not so favourable.

In sum, most of our explanatory factors are likely to be favourable insofar as the further development of participation is concerned, both in northern and southern Europe. Given all the evidence from our survey results which show not only a strong trend towards greater degrees of participation in the future in new information technology but also a high degree of consensus on many issues, we can expect that as the Community moves towards greater economic and political union following the historic agreement at Maastricht in December 1991, the benefits to both sides of industry that are so clearly evident from their experience of participation in technological change will considerably improve relations between management and employees throughout Europe.

---

1. W. W. Daniel, *Workplace Industrial Relations and Technical Change,* Frances Pinter in association with the Policy Studies Institute, London, 1987.

European Foundation for the Improvement
of Living and Working Conditions

Workplace Involvement in Technological Innovation in the European Community
Volume II: Issues of Participation

Luxembourg: Office for Official Publications of the
European Communities

1993 — 200 pp. — 16 x 23.5 cm

ISBN 92-826-5672-1 (Vol. II)

ISBN 92-826-6026-5 (Vol. I and II)

Price (excluding VAT) in Luxembourg:

Vol. II: ECU 31,50
Vol. I and II: ECU 54

Venta y suscripciones • Salg og abonnement • Verkauf und Abonnement • Πωλήσεις και συνδρομές • Sales and subscriptions • Vente et abonnements • Vendita e abbonamenti • Verkoop en abonnementen • Venda e assinaturas

BELGIQUE / BELGIË

**Moniteur belge /
Belgisch Staatsblad**
Rue de Louvain 42 / Leuvenseweg 42
B-1000 Bruxelles / B-1000 Brussel
Tél. (02) 512 00 26
Fax (02) 511 01 84

Autres distributeurs /
Overige verkooppunten

**Librairie européenne/
Europese boekhandel**
Rue de la Loi 244/Wetstraat 244
B-1040 Bruxelles / B-1040 Brussel
Tél. (02) 231 04 35
Fax (02) 735 08 60

Jean De Lannoy
Avenue du Roi 202 /Koningslaan 202
B-1060 Bruxelles / B-1060 Brussel
Tél. (02) 538 51 69
Télex 63220 UNBOOK B
Fax (02) 538 08 41

**Document delivery:
Credoc**
Rue de la Montagne 34 / Bergstraat 34
Bte 11 / Bus 11
B-1000 Bruxelles / B-1000 Brussel
Tél. (02) 511 69 41
Fax (02) 513 31 95

DANMARK

J. H. Schultz Information A/S
Herstedvang 10-12
DK-2620 Albertslund
Tlf. (45) 43 63 23 00
Fax (Sales) (45) 43 63 19 69
Fax (Management) (45) 43 63 19 49

DEUTSCHLAND

Bundesanzeiger Verlag
Breite Straße
Postfach 10 80 06
D-W-5000 Köln 1
Tel. (02 21) 20 29-0
Telex ANZEIGER BONN 8 882 595
Fax 2 02 92 78

GREECE/ΕΛΛΑΔΑ

G.C. Eleftheroudakis SA
International Bookstore
Nikis Street 4
GR-10563 Athens
Tel. (01) 322 63 23
Telex 219410 ELEF
Fax 323 98 21

ESPAÑA

Boletín Oficial del Estado
Trafalgar, 29
E-28071 Madrid
Tel. (91) 538 22 95
Fax (91) 538 23 49

Mundi-Prensa Libros, SA
Castelló, 37
E-28001 Madrid
Tel. (91) 431 33 99 (Libros)
         431 32 22 (Suscripciones)
         435 36 37 (Dirección)
Télex 49370-MPLI-E
Fax (91) 575 39 98

Sucursal:
Librería Internacional AEDOS
Consejo de Ciento, 391
E-08009 Barcelona
Tel. (93) 488 34 92
Fax (93) 487 76 59

**Llibreria de la Generalitat
de Catalunya**
Rambla dels Estudis, 118 (Palau Moja)
E-08002 Barcelona
Tel. (93) 302 68 35
       302 64 62
Fax (93) 302 12 99

FRANCE

**Journal officiel
Service des publications
des Communautés européennes**
26, rue Desaix
F-75727 Paris Cedex 15
Tél. (1) 40 58 75 00
Fax (1) 40 58 77 00

IRELAND

Government Supplies Agency
4-5 Harcourt Road
Dublin 2
Tel. (1) 61 31 11
Fax (1) 78 06 45

ITALIA

Licosa SpA
Via Duca di Calabria, 1/1
Casella postale 552
I-50125 Firenze
Tel. (055) 64 54 15
Fax 64 12 57
Telex 570466 LICOSA I

GRAND-DUCHÉ DE LUXEMBOURG

Messageries Paul Kraus
11, rue Christophe Plantin
L-2339 Luxembourg
Tél. 499 88 88
Télex 2515
Fax 499 88 84 44

NEDERLAND

SDU Overheidsinformatie
Externe Fondsen
Postbus 20014
2500 EA 's-Gravenhage
Tel. (070) 37 89 911
Fax (070) 34 75 778

PORTUGAL

Imprensa Nacional
Casa da Moeda, EP
Rua D. Francisco Manuel de Melo, 5
P-1092 Lisboa Codex
Tel. (01) 69 34 14

**Distribuidora de Livros
Bertrand, Ld.ª**
Grupo Bertrand, SA
Rua das Terras dos Vales, 4-A
Apartado 37
P-2700 Amadora Codex
Tel. (01) 49 59 050
Telex 15798 BERDIS
Fax 49 60 255

UNITED KINGDOM

HMSO Books (Agency section)
HMSO Publications Centre
51 Nine Elms Lane
London SW8 5DR
Tel. (071) 873 9090
Fax 873 8463
Telex 29 71 138

ÖSTERREICH

**Manz'sche Verlags-
und Universitätsbuchhandlung**
Kohlmarkt 16
A-1014 Wien
Tel. (0222) 531 61-0
Telex 112 500 BOX A
Fax (0222) 531 61-39

SUOMI

Akateeminen Kirjakauppa
Keskuskatu 1
PO Box 128
SF-00101 Helsinki
Tel. (0) 121 41
Fax (0) 121 44 41

NORGE

Narvesen information center
Bertrand Narvesens vei 2
PO Box 6125 Etterstad
N-0602 Oslo 6
Tel. (2) 57 33 00
Telex 79668 NIC N
Fax (2) 68 19 01

SVERIGE

BTJ
Tryck Traktorwägen 13
S-222 60 Lund
Tel. (046) 18 00 00
Fax (046) 18 01 25

SCHWEIZ / SUISSE / SVIZZERA

OSEC
Stampfenbachstraße 85
CH-8035 Zürich
Tel. (01) 365 54 49
Fax (01) 365 54 11

CESKOSLOVENSKO

NIS
Havelkova 22
13000 Praha 3
Tel. (02) 235 84 46
Fax 42-2-264775

MAGYARORSZÁG

Euro-Info-Service
Pf. 1271
H-1464 Budapest
Tel./Fax (1) 111 60 61/111 62 16

POLSKA

Business Foundation
ul. Krucza 38/42
00-512 Warszawa
Tel. (22) 21 99 93, 628-28-82
International Fax&Phone
    (0-39) 12-00-77

ROUMANIE

Euromedia
65, Strada Dionisie Lupu
70184 Bucuresti
Tel./Fax 0 12 96 46

BULGARIE

D.J.B.
59, bd Vitocha
1000 Sofia
Tel./Fax 2 810158

RUSSIA

**CCEC (Centre for Cooperation with
the European Communities)**
9, Prospekt 60-let Oktyabria
117312 Moscow
Tel. 095 135 52 87
Fax 095 420 21 44

CYPRUS

**Cyprus Chamber of Commerce and
Industry**
Chamber Building
38 Grivas Dhigenis Ave
3 Deligiorgis Street
PO Box 1455
Nicosia
Tel. (2) 449500/462312
Fax (2) 458630

TÜRKIYE

**Pres Gazete Kitap Dergi
Pazarlama Dağitim Ticaret ve sanayi
AŞ**
Narlibahçe Sokak N. 15
Istanbul-Cağaloğlu
Tel. (1) 520 92 96 - 528 55 66
Fax 520 64 57
Telex 23822 DSVO-TR

ISRAEL

ROY International
PO Box 13056
41 Mishmar Hayarden Street
Tel Aviv 61130
Tel. 3 496 108
Fax 3 544 60 39

CANADA

Renouf Publishing Co. Ltd
Mail orders — Head Office:
1294 Algoma Road
Ottawa, Ontario K1B 3W8
Tel. (613) 741 43 33
Fax (613) 741 54 39
Telex 0534783

Ottawa Store:
61 Sparks Street
Tel. (613) 238 89 85

Toronto Store:
211 Yonge Street
Tel. (416) 363 31 71

UNITED STATES OF AMERICA

UNIPUB
4611-F Assembly Drive
Lanham, MD 20706-4391
Tel. Toll Free (800) 274 4888
Fax (301) 459 0056

AUSTRALIA

Hunter Publications
58A Gipps Street
Collingwood
Victoria 3066
Tel. (3) 417 5361
Fax (3) 419 7154

JAPAN

Kinokuniya Company Ltd
17-7 Shinjuku 3-Chome
Shinjuku-ku
Tokyo 160-91
Tel. (03) 3439-0121

Journal Department
PO Box 55 Chitose
Tokyo 156
Tel. (03) 3439-0124

SINGAPORE

Legal Library Services Ltd
STK Agency
Robinson Road
PO Box 1817
Singapore 9036

AUTRES PAYS
OTHER COUNTRIES
ANDERE LÄNDER

Office des publications officielles
des Communautés européennes
2, rue Mercier
L-2985 Luxembourg
Tél. 499 28 1
Télex PUBOF LU 1324 b
Fax 48 85 73/48 68 17

10/92